# Methods in Molecular Biology

*Series Editor*
John M. Walker
School of Life and Medical Sciences
University of Hertfordshire
Hatfield, Hertfordshire, AL10 9AB, UK

For further volumes:
http://www.springer.com/series/7651

# Xylem

## Methods and Protocols

Edited by

**Miguel de Lucas**

*Department of Biosciences, Durham University, Durham, UK*

**J. Peter Etchells**

*Department of Biosciences, Durham University, Durham, UK*

Humana Press

*Editors*
Miguel de Lucas
Department of Biosciences
Durham University
Durham, UK

J. Peter Etchells
Department of Biosciences
Durham University
Durham, UK

ISSN 1064-3745 ISSN 1940-6029 (electronic)
Methods in Molecular Biology
ISBN 978-1-4939-6720-9 ISBN 978-1-4939-6722-3 (eBook)
DOI 10.1007/978-1-4939-6722-3

Library of Congress Control Number: 2016963414



Printed on acid-free paper

This Humana Press imprint is published by Springer Nature
The registered company is Springer Science+Business Media LLC
The registered company address is: 233 Spring Street, New York, NY 10013, U.S.A.

# Preface

The plant vascular system is composed of two conductive tissues, the xylem and phloem, which are responsible for the transport of water and solutes throughout the plant. Xylem cells are reinforced with thick secondary cell walls that also provide much of the plants mechanical strength and are also a source of dietary fiber. The secondary walls present in the xylem also constitute the majority of plant biomass, and consequently interest in the xylem has grown considerably as plant biomass represents a renewable source of energy. The secondary cell walls are composed predominantly of the sugar polymers cellulose and hemicellulose, and as such xylem has huge potential as a source of fermentable sugars. However, in xylem cells the β(1–4) linked glucose that constitutes the cellulose is mostly present in crystalline arrays known as microfibrils, cross-linked with hemicellulose, and surrounded by the phenolic polymer lignin thus making the sugars extremely difficult to extract from plant material, and recalcitrant to digestion. The improvement of genomics, molecular biology, microscopy, and biochemistry techniques has contributed substantially to the understanding of plant vasculature in general, and the xylem in particular, and has enabled researchers to understand and manipulate xylem differentiation and composition in order to increase xylem deposition and improve characteristics such as sugar extractability. This new understanding has come hand in hand with development of novel methods. This book aims to provide detailed accounts of the techniques used for the study and characterization of the plant vascular system, with a central focus on the xylem tissue. To do so, the book is organized in three main parts:

**Part 1: Analysis of xylem development** (Chapters 1–6)
In this part, the authors describe a range of theoretical and experimental techniques, with a special focus on versatile and informative tools. These include, for example, the use of cell cultures to induce differentiation of different xylem cell types, thus enabling multiple aspects of xylem formation such as gene regulation, cell wall composition, or apoptosis to be studied in a dynamic fashion. Consequently these techniques can provide a source of material required to address novel fundamental questions.

**Part 2: Xylem characterization though imaging techniques** (Chapters 7–13)
Xylem is challenging to study at the microscopic level, particularly with live imaging, as it is positioned deep within the plant tissue and contains a thick secondary wall. To overcome this problem, a series of improved and novel techniques, described in this part, have been developed in the last few years. These include novel imaging methods and computational analysis that now allow precise quantification and characterization of xylem properties. This set of methods is also generally applicable to studying differentiation and composition in a variety of cellular systems.

**Part 3: Analysis of the xylem composition** (Chapters 14–18)
The rich polymer diversity that constitutes the xylem secondary wall has been a focus of recent research as they are source for energy, biomaterials, and dietary fibers. The precise characterization and modification of the composition of this cell wall matrix is vital for

utilization of this material for industrial purposes. This important goal has been underpinned with advances in the study of the fundamental processes controlling the deposition of this rich structure. The methods included in Part 3 of this book have been developed to answer such applied and fundamental aims.

Target audience: Plant biochemists, Cell biologists, and Developmental biologists

***Durham, UK*** ***J. Peter Etchells***
***Durham, UK*** ***Miguel de Lucas***

# Contents

## Contributors

JAVIER AGUSTI • *Plant Sciences Department, University of Oxford, Oxford, UK*
DAN AOKI • *Graduate School of Bioagricultural Sciences, Nagoya University, Nagoya, Japan*
BENJAMIN BOLLHÖNER • *Department of Plant Physiology, Umeå Plant Science Centre, Umeå University, Umeå, Sweden*
LIAM CAMPBELL • *University of Manchester, Manchester, UK*
ANA I. CAÑO-DELGADO • *Department of Molecular Genetics, Centre for Research in Agricultural Genomics (CRAG) CSIC-IRTA-UAB-UB, Barcelona, Spain*
MARIO COIRO • *Department of Biology, ETH Zurich, Zürich, Switzerland; Department of Systematic and Evolutionary Botany, University of Zurich, Zürich, Switzerland*
RAPHAËL DECOU • *Department of Plant Physiology, Umeå Plant Science Centre (UPSC), Umeå University, Umeå, Sweden*
SACHA ESCAMEZ • *Department of Plant Physiology, Umeå Plant Science Centre, Umeå University, Umeå, Sweden*
NORMA FÀBREGAS • *Department of Molecular Genetics, Centre for Research in Agricultural Genomics (CRAG) CSIC-IRTA-UAB-UB, Barcelona, Spain*
PAU FORMOSA-JORDAN • *The Sainsbury Laboratory, University of Cambridge, Cambridge, UK*
HIROO FUKUDA • *Department of Biological Sciences, Graduate School of Science, The University of Tokyo, Tokyo, Japan*
KAZUHIKO FUKUSHIMA • *Graduate School of Bioagricultural Sciences, Nagoya University, Nagoya, Japan*
SUZANNE GERTTULA • *Forest Service, Pacific Southwest Research Station, Davis, CA, USA*
ANDRÁS GORZSÁS • *Department of Chemistry, Umeå University, Umeå, Sweden*
ANDREW GROOVER • *Forest Service, Pacific Southwest Research Station, Davis, CA, USA; Department of Plant Biology, Life Sciences Addition, University of California, Davis, CA, USA*
MARTA IBAÑES • *Department of Condensed Matter Physics, University of Barcelona, Barcelona, Spain*
YUKI KONDO • *Department of Biological Sciences, Graduate School of Science, The University of Tokyo, Tokyo, Japan*
MANOJ KUMAR • *University of Manchester, Manchester, UK*
CHARLES W. MELNYK • *The Sainsbury Laboratory, University of Cambridge, Cambridge, UK*
DELPHINE MÉNARD • *Department of Plant Physiology, Umeå Plant Science Centre (UPSC), Umeå University, Umeå, Sweden; Department of Ecology, Arrhenius laboratories, Environment and Plant Sciences (DEEP), Stockholm University, Stockholm, Sweden*
JENNIFER C. MORTIMER • *Lawrence Berkeley National Laboratory, Joint BioEnergy Institute, Emeryville, CA, USA*
ALIF MEEM NURANI • *Department of Biological Sciences, Graduate School of Science, The University of Tokyo, Tokyo, Japan*
YOSHIHISA ODA • *Center for Frontier Research, National Institute of Genetics, Mishima, Shizuoka, Japan; Department of Genetics, SOKENDAI (Graduate University for Advanced Studies), Mishima, Shizuoka, Japan*

EDOUARD PESQUET • *Department of Plant Physiology, Umeå Plant Science Centre (UPSC), Umeå University, Umeå, Sweden; Department of Ecology, Arrhenius laboratories, Environment and Plant Sciences (DEEP), Stockholm University, Stockholm, Sweden*
LAURA RAGNI • *Center for Plant Molecular Biology – ZMBP, Developmental Genetics, University of Tübingen, Tübingen, Germany*
PIERRE BARBIER DE REUILLE • *University of Bern, Bern, Switzerland*
JORGE REYES-RIVERA • *Departamento de Botánica, Instituto de Biología, Universidad Nacional Autónoma de México, Mexico*
MASATO SAITO • *Department of Biological Sciences, Graduate School of Science, The University of Tokyo, Tokyo, Japan*
HENRIK SERK • *Department of Plant Physiology, Umeå Plant Science Centre (UPSC), Umeå University, Umeå, Sweden; Department of Medical Biochemistry and Biophysics, Umeå University, Umeå, Sweden*
TERESA TERRAZAS • *Instituto de Biología, Universidad Nacional Autónoma de México, Mexico*
ELISABETH TRUERNIT • *Department of Biology, ETH Zurich, Zürich, Switzerland*
HANNELE TUOMINEN • *Department of Plant Physiology, Umeå Plant Science Centre, Umeå University, Umeå, Sweden*
SIMON TURNER • *University of Manchester, Manchester, UK*
RAYMOND WIGHTMAN • *Microscopy Core Facility, Sainsbury Laboratory, University of Cambridge, Cambridge, UK*
BAOCAI ZHANG • *State Key Laboratory of Plant Genomics, Institute of Genetics and Developmental Biology, Chinese Academy of Sciences, Beijing, China*
YIHUA ZHOU • *State Key Laboratory of Plant Genomics, Institute of Genetics and Developmental Biology, Chinese Academy of Sciences, Beijing, China*

# Part I

## Development

# Chapter 1

# Experimental and Theoretical Methods to Approach the Study of Vascular Patterning in the Plant Shoot


## Norma Fàbregas, Pau Formosa-Jordan, Marta Ibañes, and Ana I. Caño-Delgado



## Abstract

The plant vascular system provides transport and mechanical support functions that are essential for suitable plant growth and development. In *Arabidopsis thaliana* (Arabidopsis), the vascular tissues at the shoot inflorescence stems are disposed in organized vascular bundles. The vascular patterning emergence and development within the shoot inflorescence stems is under the control of plant growth regulators (De Rybel et al., Nat Rev Mol Cell Biol 17:30–40, 2016; Caño-Delgado et al., Annu Rev Cell Dev Biol 26:605–637, 2010). By using a combined approach of experimental methods for vascular tissues visualization and quantification together with theoretical methods through mathematical and computational modeling, we have reported that auxin transport and brassinosteroid signaling play complementary roles in the formation of the periodic vascular patterning in the shoot (Ibañes et al., Proc Natl Acad Sci U S A 106:13630–13635, 2009; Fàbregas et al., Plant Signal Behav 5:903–906, 2010; Fàbregas et al., PLoS Genet 11:e1005183, 2015). Here, we report the methodology for the interdisciplinary analysis of the shoot vascular patterning in the plant model Arabidopsis into a handle procedure for visualization, quantification, data analysis, and modeling implementation.



**Key words** Arabidopsis, Modeling, Vascular patterning, Brassinosteroids, Auxin, Shoot, Influx carriers, Efflux carriers, Cell division, Procambium


## 1 Introduction

The vascular tissues in *Arabidopsis thaliana* (Arabidopsis) plants are primarily composed by the xylem, which transports water, and the phloem, which transports solutes and plant photo-assimilates. In Arabidopsis shoots, the xylem and phloem tissues are periodically disposed in vascular bundles forming a radial pattern, whereas the root vasculature displays a diarch pattern, having two phloem poles at both sides of a single xylem axis [1–4]. When viewed in cross sections, Arabidopsis root vascular tissues display two phloem

---

Norma Fàbregas and Pau Formosa-Jordan contributed equally to this work


Miguel de Lucas and J. Peter Etchells (eds.), *Xylem: Methods and Protocols*, Methods in Molecular Biology, vol. 1544, DOI 10.1007/978-1-4939-6722-3_1, © Springer Science+Business Media LLC 2017

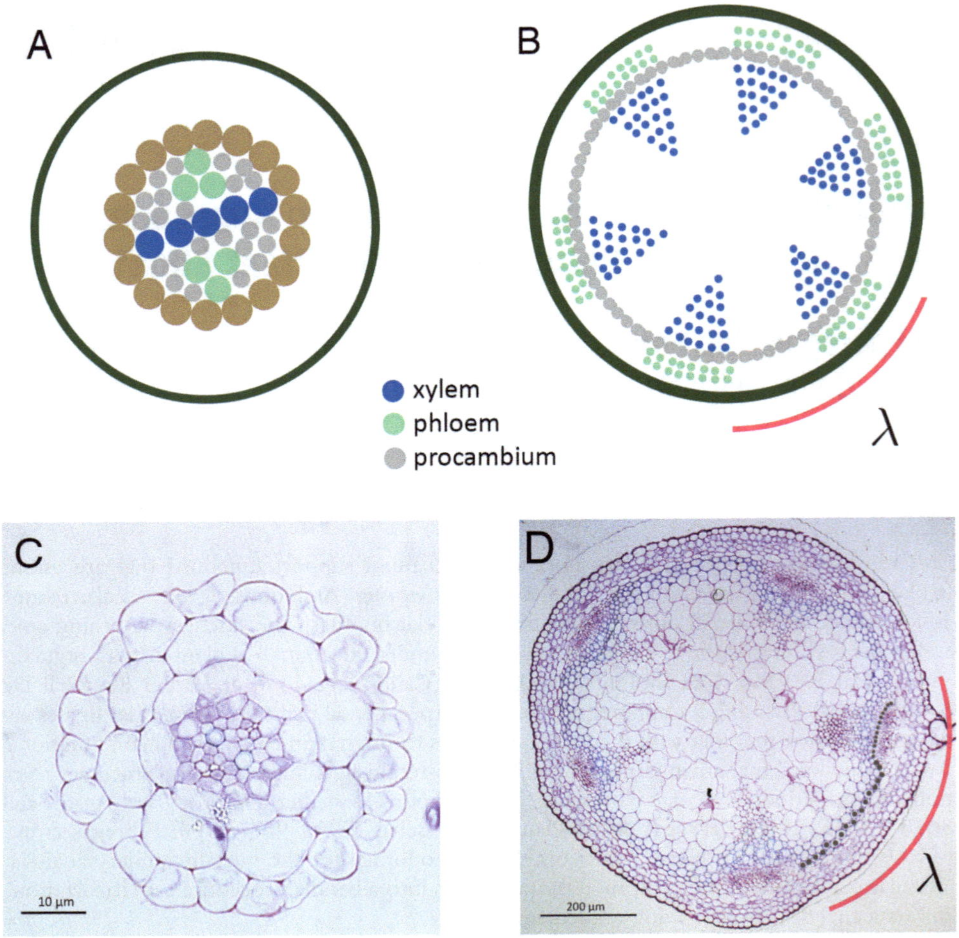


**Fig. 1** Schematic representation of the root and shoot to depict the different vascular pattern organizations in Arabidopsis. (**a**) Diarchal vascular patterning in the Arabidopsis primary root schematic cartoon and (**c**) anatomical section stained with toluidine blue. (**b**) Radial section of the periodical vascular pattering in the Arabidopsis shoot inflorescence stems schematic cartoon and (**d**) anatomical section stained with toluidine blue. Procambial cells are depicted in *grey*, phloem cells in *green* and xylem cells in appear in *blue* (**a**, **b**). VB unit or wavelength is formed by the number of procambial cells (depicted in *grey*) within the VB spacing and represented in *pink* (λ)

poles (Fig. 1a, green cells) diametrically opposed and a central xylem cells axis (Fig. 1a, blue cells) forming a diarchy pattern, and surrounded by procambial meristematic cells (Fig. 1a, grey cells). Instead, at shoot inflorescence stems the vascular bundles (VB) are periodically disposed along a circular ring of pro-vascular cells. Each VB is composed by a group of meristematic procambial cells (Fig. 1b, grey cells), differentiated phloem cells (Fig. 1b, green cells) and differentiated xylem cells (Fig. 1b, blue cells), together forming a collateral pattern [3–5]. In between the VBs, interfascicular fibers (IF) differentiate, supporting the inflorescence stem (Fig. 1b, grey cells between VBs [3, 4]). While numerous studies have approached the root vascular patterning and development in

Arabidopsis [6–11], the shoot vascular tissue patterning has been less understood due to the tissue organization complexity and the high variability within the shoot anatomy.

Additional limitations in the study of the shoot vascular patterning are associated to inner location of the vascular tissues, which hampers the use of microscopy techniques. The vascular pattern depends on auxin transport [12–14], which is actively directed by the so-called efflux and the influx carriers [15–17]. Our previous analysis proposed that the periodic pattern arises through the accumulation of the phytohormone auxin and that brassinosteroids control the number of procambial cell divisions within the VBs in the shoot inflorescence stem [12, 38]. In addition, we have recently reported that auxin influx carriers modulate the number of VBs and xylem differentiation [18]. In both studies, we benefit from the combination of theoretical and experimental approaches to unravel the role of the auxin transport and brassinosteroids signaling in the formation of plant vasculature in the shoot.

Herein, based on our previous studies [12, 18, 38] we describe methods combining experimental and theoretical approaches for investigating shoot vascular patterning and development. In short, Arabidopsis shoot inflorescence stems are embedded in Historesin and subsequently sectioned for microscopy analysis. Imaging cross stem sections enables an exhaustive quantification of specific cell and vascular pattern related parameters. In the following paragraphs, we describe in detail how to obtain the vascular parameters data and how to implement a mathematical and computational model on auxin transport in a simplified geometry for studying VB pattern formation. Finally, we discuss the characterization of the simulation results and the comparison with the experimental data. The integration of quantitative biological data with mathematical and computational modeling permits to build up new hypothesis that can be subsequently tested experimentally, leading to the formulation of new models for the vascular pattern development.

## 2 Materials

### 2.1 *Plant Material*

1. Sterilization solution: 35% of hypochlorite solution (NaClO 15% v/v) in milliQ sterile water.
2. Autoclaved milliQ.
3. 1.5 mL micro centrifuge tubes.
4. Squared petri dishes (10 × 10 cm).
5. Micropore film.
6. Growth media: MS (Murashige and Skoog salt media supplemented with vitamins, 10 g/L of sucrose and 5 mM MES/KOH, pH 5.7) (*see* **Note 1**).
7. Laminar flow clean bench.

8. Autoclaved soil mixture supplemented with 8:1:1 perlite and 8:1 vermiculite for aeration.
9. Plastic trays and plastic covers.
10. Growth chamber set up for long day or short day conditions (*see* **Note 2**).

### 2.2 Fixation, Dehydration and Embedding

1. Razor blades.
2. Glass vials (e.g., 20 mL) with screw caps.
3. 1.25% glutaraldehyde in 0.1 M sodium cacodylate buffer (pH 7.4).
4. Ethanol dilution series (30%, 50%, 70%, 90%, and 100%).
5. Historesin-I: 100 mL Historesin solution + 1 g hardener I (Technovit 7100, Heareus Kulzer, Electron Microcopy Science).
6. Historesin-II: 15 mL Historesin-I + 1 mL hardener II (Technovit 7100, Heareus Kulzer, Electron Microscopy Science) (*see* **Note 3**).
7. Flat polystyrene tray filled with ice.
8. Plastic molds (e.g., Electron Microcopy Science).
9. Tweezers.
10. Parafilm.

### 2.3 Ultramicrotome Sectioning

1. Microtome (e.g., RM2265 microtome, Leica).
2. Punchs.
3. Paint brush.
4. Slides (76 × 26 mm).
5. Razor blades.
6. Glass razor blades (*see* **Note 4**).
7. Glass Knife Maker (e.g., EM KMR3, Leica) (*see* **Note 4**).
8. Hot plate.

### 2.4 Staining and Imaging Acquisition

1. Staining solution: 0.1% toluidine blue in 0.1 M $NaHPO_3$, pH 7.0.
2. Coplin staining jars.
3. Water.
4. Coverslips.
5. Ethanol 70%.
6. Axiophot Microscope (e.g., Zeiss).

### 2.5 Quantitative and Statistical Vascular Parameters Analysis

1. Computer.
2. ImageJ software (http://rsb.info.nih.gov/ij/).
3. Excel or equivalent.
4. Sigma Plot, R or equivalent.

*2.6 Mathematical and Computational Modeling*

1. Computer.
2. Code or software for performing simulations, e.g.,
   (a) Mathematica notebook provided in [18] (S1_Code.nb in Supplementary Material).
   (b) More advanced modelers should use "Organism" package, developed by Prof. Jönsson's Lab, which has been used for studying plant morphogenesis (*see* http://dev.thep.lu.se/organism/ and *see* **Note 5**).
3. Code or software for data analysis and representation (e.g., the ones suggested in **item 4** of Subheading 2.5, Mathematica, Matlab, or Python).

## 3 Methods

*3.1 Plant Material and Growth Conditions*

1. Sterilize seeds by adding 35 % of hypochlorite solution (NaClO 15% v/v) in 1.5 mL microcentrifuge tubes containing the seeds and shake for 5 min (*see* **Note 6**).
2. Rinse with sterile milliQ water for 5 min (×5).
3. Place the seeds at 4 °C for 48 h to synchronize germination.
4. Plate seeds on petri dishes containing 1× Murashige and Skoog (MS) medium and seal with micropore tape.
5. Place the plates in a growth chamber with long day photoperiod conditions 16 h light/8 h dark; 8470 lux; 20–23 °C) or short day photoperiod (8 h light/16 h dark; 8470 lux; 20–23 °C) (*see* **Note 2**).
6. Prepare individual plastic pots with soil mixture, place them in a plastic tray and water them until saturation.
7. Transfer 10-day-old seedlings carefully from the MS plates to the soil, hiding the roots deep in the soil.
8. Cover the plastic trays with plastic lids or cling film and place the plant trays in the growth chamber under long day or short day photoperiod conditions (*see* **Note 2**).
9. After 48 h, perforate the cover film to slowly acclimate plants to environmental humidity.
10. After 2 days, remove the covering plastic completely and start watering plants normally.

*3.2 Material Collection, Fixation and Embedding*

1. Grow up to 5-week-old plants in long-day conditions or 14-week-old plants in short-day conditions (*see* **Note 7**).
2. Cut the entire main shoot inflorescence stem approximately 1 cm above the rosette. Then, discard the rosette and section 1 cm of basal zone of the inflorescence stem, where the vascular tissues are more differentiated. Discard the rest of the shoot

(*see* **Note 8**). Notice that the number of VB does not change significantly along the apica-basal axis [12]. However, the number of cells does decreases with increasing proximity to the apical meristem and differentiation of the vascular cells dramatically reduces as well [12]. Unless using a fluorescent marker for identifying the vasculature, we highly recommend sectioning approximately 1 cm from the basal stem above the rosette, where the vascular pattern is well established and differentiated, and VBs can be seen directly under the microscope with white light.

3. Fix inflorescence stem sections in 1.25% glutaraldehyde in 0.1 M sodium cacodylate buffer (pH 7.4), infiltrate the samples in vacuum during 30 min, and keep them at 4 °C overnight. Samples can be kept for several months at this step.
4. Remove the fixation solution and dehydrate the samples through a graded ethanol series: 30%, 50%, 70%, 90%, and 100% for 60 min each and shaking at RT.
5. Remove ethanol and start Historesin-I infiltration by adding Historesin-I–ethanol solution (1:1). Incubate at RT for 60 min and constant agitation.
6. Incubate samples with Historesin-I 100% at RT for 60 and constant agitation. Repeat this step.
7. Add newly fresh prepared Historesin-I 100% solution and keep samples at 4 °C overnight.
8. Label plastic molds with each sample name and place them on ice, in horizontal position (*see* **Note 9**).
9. Place one shoot stem inflorescence sample (0.5 cm approx.) into each plastic mold and fill it with freshly prepared 100% Historesin-II solution.
10. Distribute and orientate each inflorescence stem sample lengthwise the mold carefully, using tweezers.
11. Carefully prepare and cut Parafilm pieces to completely cover the area of each individual mold, place it on the top of your sample and keep them overnight at 4 °C to accelerate the Historesin polymerization (*see* **Note 10**).
12. Sample blocks can be stored at RT for months until section.

### 3.3 Microtome and Imaging

1. Remove solidified block from the plastic molds by prying with a punch.
2. Trim away excess of Historesin exposing the sample in a pyramidal mode and fix the block on the microtome holder for transverse sectioning.
3. Place a glass razor blade in the appropriate orientation and cut Historesin away (trimming mode: 5–10 μm) until reaching the inflorescence stem sample (*see* **Note 11**).

4. Once sample is reached, cut numerous sections per sample, selecting 3 μm of thickness and the cut mode of your choice in the Microtome (e.g., rock mode, *see* **Note 12**).
5. Label each slide and place a drop of milliQ water on the top of the slide.
6. Collect several tissue sections with a punch or brush and place them on the slide.
7. Evaporate water from the samples by placing the slide on a hot plate for 10 min at 80 °C.
8. Stain sections by submerging slides with 0.1 % Toluidine blue in 0.1 M $NaPO_4$ pH 7.0.
9. Rinse slides with sterilized water and clean over staining in the backside of the slide with EtOH 70 % for optimal imaging.
10. Add a drop of water and a coverslip.
11. Acquire images with an Axiophot Microscope (Zeiss) under white light (set the zoom at 20× for a global imaging of the shoot stem sample and at 40× for a higher resolution of each cell-type imaging within the same sample).

### 3.4 Quantitative Vascular Parameters and Statistical Analysis

Different vascular features or parameters can be extracted from the microscopy images. Some of them might be more appropriate to get in metric units (μm) and others might be useful in 'number of cells' units (where 'number of cells' units would not have a strict conversion into metric units, given that cells can have different sizes). We have shown that the VB number in brassinosteroids gain-of-function and loss-of-function mutants did not correlate with the pro-vascular ring length when measured in metric units, but it did change when measured in 'number of cells' units [12]. We interpreted this feature as there is a pattern characteristic length in cells number, but not in metric units, that would provide the spacing between VBs. Hence, it is worth taking into account not just metric units but also number of cells for VBs characterization.

1. Manually count the VB number from microscope images (*see* **Note 13**).
2. Manually count the number of cells for each vascular unit, which is formed by the procambial cells within a VB plus the clockwise adjacent interfascicular fiber cells (Fig. 1b and d, grey cells). Sum of all these cells per plant section, what gives the number of cells along the provascular ring, $N$.
3. Measure average stem diameter by measuring two orthogonal diameters and interfascicular fibers length for each sample with the ImageJ software (http://rsb.info.nih.gov/ij/).
4. Compute all the average values for each parameter, obtaining a single (averaged) value of each quantity for each plant. Generate data sets of the averaged quantities for a set of $n$ plants. The

average value of the number of cells per vascular unit constitute the measure of the VB spacing, and due to its regularity, its average can also be called as a characteristic wavelength of the VB pattern, $\lambda$ (Fig. 1b).

5. Run Wilcoxon Mann–Whitney rank sum statistical test (in Matlab) on your data sets (samples of size $n$) to evaluate whether the data from the mutant or treated samples are significantly statistically different with respect to the non treated WT (*see* **Note 14**).
6. Differences in VB units between the WT and a certain mutant or treatment can be due to a change of VB spacing ($\lambda$) and the total number of cells along the provascular ring ($N$). Quantification of the contribution of these two factors to the change in VB number ($V$) can be computed through

$$\frac{\Delta V}{V_{\mathrm{WT}}} = \frac{\Delta \lambda}{\lambda_{\mathrm{WT}}} + \frac{\Delta N}{N_{\mathrm{WT}}}, \tag{1}$$

where $\Delta X = X_{\mathrm{mutant}} - X_{\mathrm{WT}}$ with $X$ being the median value found for each variable in a sample of plants. This relation stems from $V = N/\lambda$. The percentage of contribution of VB spacing is then $100\Delta\lambda V_{\mathrm{WT}}/(\Delta V \lambda_{\mathrm{WT}})$.

### 3.5 Mathematical Modeling

We used a model proposed for phyllotactic patterning [19] for studying how auxin maxima can be formed across a provascular ring of cells leading to the formation of the VBs. The modeling readership will have to refer to other articles in the field for having a view on the different models for auxin transport and the recent advances in this direction (e.g., [20–24]).

1. To model the periodic arrangement of VBs in the shoot and take into account the symmetry of the provascular ring, model a circular ring of cells and the apoplast in between them by using a single linear chain of cells (and apoplast) with periodic boundary conditions (Fig. 2a, *see* **Note 15**).
2. Write down the ordinary differential equations to model the auxin transport dynamics inside cells and in the apoplastic space between cells. Since auxin is quickly diffusing in the cytosol [25], homogeneous auxin distribution is considered inside each cell. We define as $Ai$ the concentration of auxin inside cell $i$, and $ai$ the concentration of auxin in the apoplast $i$ in between cell $i$ and cell $i+1$ (*see* **Note 16**).

   Assume auxin is constantly produced and linearly degraded inside cells, and that is actively and passively transported from cells to the apoplast and vice versa. Assume transport between apoplastic spaces.

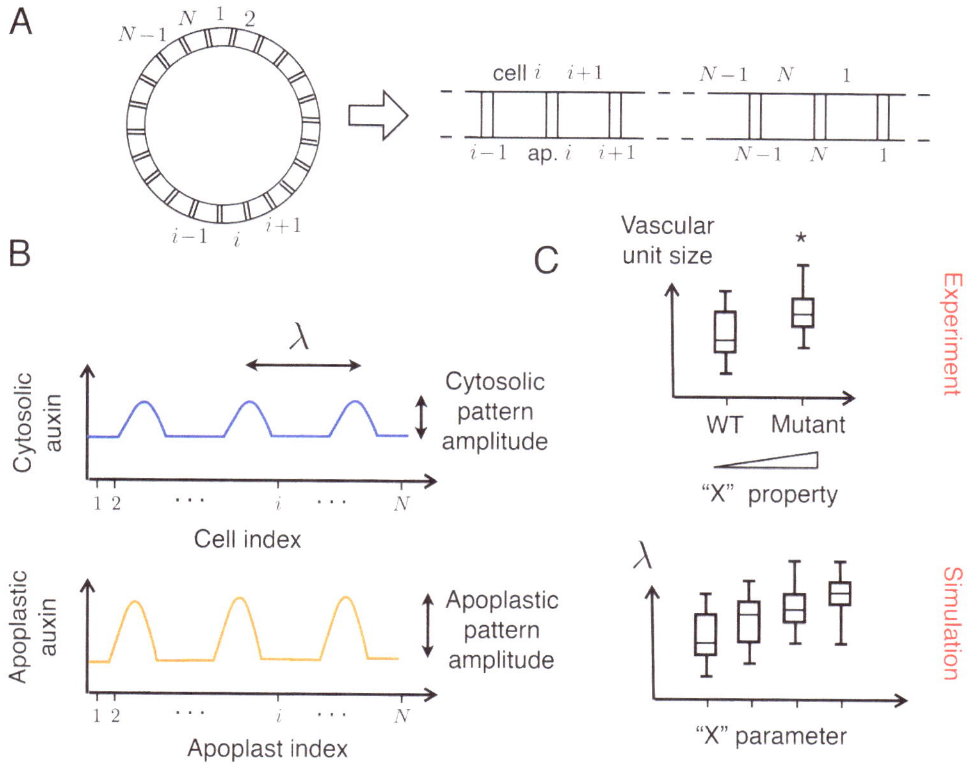


**Fig. 2** Modeling vascular pattern formation in the shoot. (**a**) Schematic representation of the provascular ring made of cells (*bigger compartments*) and apoplasts (*smaller compartments*) (*left*). This is equivalent to model a chain of cells and apoplast compartments with periodic boundary conditions (*right*). In the ring, *labels* denote cell indexes. In the chain of cells, *labels* denote cell indexes (*top*) and apoplast indexes (*bottom*). (**b**) Cartoon of a simulation result of a periodic pattern of cytosolic and apoplastic auxin concentration peaks. Cytosolic and apoplastic patterns have the same wavelength or peak-to-peak spacing ($\lambda$), but can have different amplitudes and peak widths (normally we would refer to $\lambda$ as the average peak-to-peak distance, although in the panel we indicate a single peak-to-peak distance for simplicity). (**c**) Cartoon of a comparison between experimental results (*top*) and simulation results (*bottom*) of the pattern wavelength (average vascular unit per plant in the experiments, and average peak-to-peak spacing along a simulated provascular ring). Boxplots are used for representing the measured wavelength in a sample of plants and in multiple realizations of the simulation with different random initial conditions. In the experiments, we see that mutant plants enhancing a certain "X" property (e.g., an effective parameter related to auxin transport) enlarges the vascular unit size in a significant manner (* denotes a $p$-value $\leq 0.01$). This trend is reproduced in simulations; when the parameter "X" is increased, the wavelength $\lambda$ is also enhanced

(a) The model for the dynamics of auxin inside cell $i$ reads

$$\frac{dA_i}{dt} = -\sum_{j\in n(i)} W_{ij} J_{ij} - \nu_c A_i + \sigma_c, \tag{2}$$

where $\sigma_c$ and $\nu_c$ are the auxin production and degradation rates, and $Jij$ stands for the auxin flux from the cell $i$ to the apoplast $j$ and contains the protonated passive auxin transport. $Wij$ is the ratio between the area of the cell $i$ facing the apoplast $j$ divided by the volume of the cell $i$. $n(i)$ refers to the indices of the apoplastic compartments neighboring cell $i$ (Fig. 2). Auxin concentration in the apoplast compartment $i$ has the dynamics

$$\frac{da_i}{dt} = \sum_{j\in N(i)} W_{ji} \frac{V_{\text{cell}}}{V_{\text{ap}}} J_{ji} + D_{\text{w}} \nabla_i^2 a_i, \tag{3}$$

where $D_{\text{w}}$ would be an effective auxin diffusion rate between the apoplastic compartments, $V_{\text{cell}}$ would be the cell volume and $V_{\text{ap}}$ the apoplast volume. $N(i)$ refers to the indices of the cells neighboring apoplast $i$. $\nabla i^2$ is the discrete Laplacian in 1D, which takes the form in this geometry of $\nabla i^2\, ai = (ai_{+1} + ai_{-1} - 2ai)/\Delta L^2$, being $\Delta L$ the space between two apoplastic spaces.

(b) Write down the different terms contributing to the auxin flux from cell $i$ to compartment $j$. Consider $\omega i$ the protonated passive auxin transport from cell $i$ to the apoplast $j$, $\alpha i$ the active auxin influx to cell $i$ due to the auxin influx carriers, and $\varphi i$ the active auxin efflux from cell $i$ due to the auxin PIN transporters, so that the flux reads

$$J_{ij} = \omega_i - \alpha_i + \varphi_i. \tag{4}$$

The precise formulation of these terms can be found in [18, 19] (*see* **Note 17** [19]).

(c) The active efflux $\varphi i$ requires further attention, given that the PIN transporters are polarly expressed in the cell and it has been shown that this is fundamental for driving the auxin maxima and, ultimately, vascular pattern formation [12, 14]. In our current formulation, we assume the up-the-gradient hypothesis [26, 27], where PINs tend to localize in the membrane that faces the neighbor cell with higher auxin concentration (*see* **Note 17**). This leads PINs converging towards certain regions in the tissue, driving auxin maxima formation. Though, other PIN polarity mechanisms could be considered for vascular patterning formation [20–24].

(d) In order to evaluate the patterning resulting for different parameter values, it might be worth defining effective parameters, defined as functions of the former parameters. The way these effective parameters can be defined will naturally appear from rearranging and grouping the parameters in the formulated equations. Performing this reduction will greatly reduce the number of parameters to be explored. Some pattern properties can be better understood with respect to these effective parameters, rather than from the individual parameters per se initially defined. Then, write the dynamic equations in terms of these effective parameters (*see* **Note 18**, and S1 Text of [18]).

3. Assign parameter values to the model. Some parameters have already been measured or estimated [26, 28–31]. Nevertheless, the model should not show high sensitivity to parameter changes.
4. Assign initial conditions of the different variables. One option is to provide small random fluctuations around the homogeneous state of the dynamics. The homogeneous state can be found analytically or numerically (e.g., by using NSolve in Mathematica, *see* S1 Code in [18]). The VBs pattern initiation should consider an initial auxin pre-pattern of peaks, emulating the existence of certain preformed vascular veins (e.g., the ones that might be created by the cotyledons [32]).
5. To obtain the time evolution of the auxin concentrations, one needs to make numerical simulations of the described equations. One can use "Mathematica", with the "NDSolve" function with a "StiffnessSwitching" method to avoid numerical instabilities, or "Matlab" software, with a solver like "ode45" or "ode15s" (*see* **Note 19**). One should establish a criterion to stop the simulation at a certain time step; otherwise, a steady state criterion might lead to very long simulations due to coarsening dynamics.
6. For computing the patterning spacing, count the number of cytosolic auxin maxima and divide it by the number of cells along the simulated provascular ring. The inverse of this quantity is the average number of cells between two auxin maxima or pattern wavelength $\lambda$, which is a proxy for the number of cells per vascular unit (*see* **Notes 20** and **21**).
7. Other observables from the auxin profile can be extracted, as the cytosolic and apoplastic average auxin peak maxima, the peak amplitudes (mean difference between maxima and auxin minima, Fig. 2b), the average auxin concentration in the cells, or time for pattern emergency. Some of these observables might still be compared with the experiments (Fig. 2c). Amplitudes of the cytosolic auxin pattern might be a proxy for understanding the measurements of the new quantitative auxin sensor R2D2 [33] across shoot sections.
8. One way to formulate a hypothesis to be tested experimentally is to study how the different characterized observables of the simulated pattern change with respect to a certain model parameter (Fig. 2c). We previously characterized how the distance between peaks (VB unit size) changed with respect to influx levels [18]. Giving a quantitative model prediction of the precise value of the peak spacing might be very difficult, if not impossible, due to the model simplifications, and to the unknown parameter values. But understanding how the pattern is modulated due to a certain kind of parameter change,

this is something that can be tested by comparing the patterning properties of the control WT plants and the plants in which this parameter is experimentally changed (*see* **Notes 21–23**).

9. One may want to change the model in the present formulation to test a specific hypothesis of how a particular gene can modify a pattern. For instance, one might want to test how a specific gene $X$ that enhances auxin production and at the same time is enhanced or inhibited by auxin itself would change the pattern. To do so, one can model the gene dynamics in cell $i$ as

$$\frac{dX_i}{dt} = \alpha + f(A_i) - \beta X_i, \tag{5}$$

where $\alpha$ would be a basal production rate and $\beta$ the degradation rate for such gene, and $f(Ai)$ would be a Hill-like increasing or decreasing function for denoting activation or inhibition (*see* **Note 24**). Auxin dynamics would read now

$$\frac{dA_i}{dt} = -\sum_{j\in n(i)} W_{ij} J_{ij} - v_c A_i + \sigma_c + \sigma_c' X_i, \tag{6}$$

where the last term would be the auxin dependent production term, being $\sigma'$ a production rate.

## 4 Notes

1. Only after adjusting the pH of MS media up to pH 5.7 with KOH, add plant agar and autoclave the bottle media.
2. Long day conditions are standardly used for growing plants while some plant and vascular development phenotypes are more solid when plants are grown in short day conditions. For instance, WT plants show much more xylem differentiation and a significantly higher number of vascular bundles growing in short day conditions when compared to long day conditions [18].
3. Historesin I and II are very toxic. Take safety measures to handle them, specifications are well described in Technovit® instruction manual. Historesin II polymerizes very fast, prepare small volume that you will use in few minutes (10 mL approx).
4. Glass razor blades are prepared by scratching a glass bar with a diamond by the half and then pressure is applied in the center of the line. For a detailed video description of Leica Glass Knife Maker operation see https://www.youtube.com/watch?v=nXF1Mje_fNU.
5. In Organism package, cell growth, division, and mechanical cues can also be implemented [26, 34]. This package enables the implementation of more sophisticated 2D and 3D

geometries of cellular arrays, and it also allows the implementation of cell wall compartments.

6. All mutant plants analyzed were in *Arabidopsis thaliana* Columbia-0 (Col-0) ecotype background. Take in account that different backgrounds are not comparable at the phenotypic level, especially in those related to vascular development and differentiation.
7. In long day conditions plants should be grown until 5 weeks (bolting normally occur around 3-week-old plants or 21DAG) and in short day conditions plants should be grown until 10 weeks (depending on mutant phenotypes related to light or flowering this time can be extended up to 14 weeks).
8. Short day conditions can occasionally drive the emergence of aerial rosettes in both WT and mutant adult plants. Plants showing those aerial rosettes development were not considered for our analysis.
9. Historesin solution can easily erase the labeling even when you are using a permanent marker.
10. Make sure that the Parafilm completely covers the sample and avoid any air bubbles inside the block since it will not permit a homogeneous and proper polymerization of the Historesin (The polymerization of the Historesin at 4 °C is oxygen-sensitive).
11. Handle with care. Never touch the glass razor blade cutting edge with any other material other than a paintbrush, which is useful to clean hisotresin leftovers away. Any other direct physical contact with the razor blade cut edge, including your hands, will damage it.
12. Microtome specifications can be downloaded from the instructions manual depending on the microtome brand.
13. Rarely, fusions of two VB can be observed within the samples. It is important to establish a criterion for considering one or two VBs when duplications are found. Distinct VB fusions scenarios can be observed: two phloem peaks in one xylem peak, two phloem peaks in two xylem peaks, one phloem peak on two xylem peaks. We considered the first two scenarios with separated phloem units to be independent VBs while the last one where only one phloem pole appeared, it was considered one VB (duplication).
14. This test is applied for non-normally distributed data, and evaluates whether the data sets are showing a significant statistical difference by comparing its medians. When ties were present the test results were checked by performing a permutation test (using 9999 permutations).

15. For simplicity, here we omit cell division, but the reader can refer to [12] to see how it can be implemented in a provascular ring. VB pattern formation may become a 3D problem in case we want to take into account the possible effect of the phyllotactic pattern or the phyllotactic strands along the shoot [35]. Modeling the pattering process in 3D becomes a much more complex problem, and this is not in the scope of this chapter.
16. Both the cells and the apoplast are described as a single point each (transport inside cells or within the apoplast in between two neighboring cells is not modeled). Apoplastic compartments are modeled between cells, and not surrounding the whole cell (Fig. 2a). A fundamental advantage of the proposed geometry and these assumptions is that it facilitates further analytical work on the model (*see* **Note 21**). If one considered the apoplastic spaces surrounding the whole cell, the transport between apoplastic compartments and between apoplast and cellular compartments should be formulated differently.
17. In the present modeling formulation, transporter dynamics is assumed to be quicker than auxin dynamics, and the explicit dynamics of influx and efflux transporters variables is omitted. You can see how other models explicitly model its dynamics in [19, 26]. Note that transporters can be considered constant in the cell or upregulated by auxin itself [18, 19, 26]. The auxin flux due to these transporters can be considered linear as a first approximation.
18. Note these effective parameters may be dimensional or dimensionless, depending on the choice of the modeler and the particular questions that are being addressed. For instance, one might define $\sigma = \sigma_c / \nu_c$ as an effective production rate, what will have units of concentration instead of units of concentration divided by time. Also, one may want to do these effective parameter definitions in a non-dimensionalized system of equations, in which the variables and the time are dimensionless. This will further reduce the number of parameters. See [19] or the S1 Text in [18] as examples of non-dimensionalizations and effective parameters definitions.
19. Note that in some regions of the parameter space, especially when coefficients of the different contributing terms are very different in magnitude, there might be numerical instabilities causing numerical artifacts. In these cases, be sure that the integrator is using a sufficiently small time step (e.g., by using a fourth order Runge-Kutta algorithm [36] with a sufficiently small time step), or chose an algorithm more suitable for what is called stiff dynamical systems.
20. Sometimes, small incipient peaks emerge along the pattern, and you may do not want to count them into the quantification. One can develop a criterion to neglect them discarding

the peaks with smaller amplitudes than 0.15 the amplitude of the average peak amplitude in a simulation.

21. In some models, for certain kind of regular geometries as the one we have chosen, a theoretical prediction of the patterning spacing can be performed through the use of analytical calculations. In particular, linear stability analysis can show how the patterning spacing can change due to different parameters (*see* [19, 26] and S1 Text in [18] for the presented model, and *see* [37] for more general considerations).
22. Mathematica offers the "Manipulate" option, which is very useful to study pattern changes upon changing one of the model parameters.
23. At each point of the parameter space different repetitions should be done, with different randomness in the initial conditions, emulating the quantification of different plants. Since the quantified observables might not follow a Gaussian distribution, a boxplot representation of them for the different parameter values might be useful (*see* [18] and Fig. 2c).
24. An increasing Hill function of auxin *Ai* can be defined as

$$f(A_i) = \frac{V\,A_i^n}{K^n + A_i^n}, \qquad (7)$$

while a decreasing function would be

$$f(A_i) = \frac{V}{1 + \left(\frac{A_i}{K}\right)^n}, \qquad (8)$$

being $V$ the maximal production term, $K$ the auxin level at which $f(Ai)$ has its maximal value, and $n$ the exponent denoting more or less nonlinearity in such auxin-mediated regulation.

## Acknowledgments

We acknowledge financial support from the Spanish Ministry of Economy and Competitiveness, through the "Severo Ochoa Programme for Centres of Excellence in R&D" 2016–2019 (SEV-2015-0533)". N.F. is funded by "Fundación RENTA CORPORACIÓN" charity in A.C.-D. Lab. P.F.-J. acknowledges the postdoctoral fellowship provided by the Herchel Smith Foundation. P.F.-J and M.I. acknowledge support from the Spanish Ministry of Economy and Competitiveness through grants FIS2012-37655-C02-02 and FIS2015-66503-C3-3-P and to the Generalitat de Catalunya through Projecte Consolidat 2014 SGR

878. A.I.C.-D. lab is funded by a BIO2013-43873 grant from the Spanish Ministry of Economy and Competitiveness, and the European Research Council by the ERC Consolidator Grant (ERC-2015-CoG—683163).

## References


1. De Rybel B, Mahonen AP, Helariutta Y, Weijers D (2016) Plant vascular development: from early specification to differentiation. Nat Rev Mol Cell Biol 17(1):30–40. doi:10.1038/nrm.2015.6
2. Caño-Delgado A, Lee JY, Demura T (2010) Regulatory mechanisms for specification and patterning of plant vascular tissues. Annu Rev Cell Dev Biol 26:605–637. doi:10.1146/annurev-cellbio-100109-104107
3. Essau K (1977) Anatomy of seed plants, 2nd edn. Wiley, New York
4. Essau K (1965) Plant anatomy, 2nd edn. Wiley, New York
5. Jurgens G (2001) Apical-basal pattern formation in Arabidopsis embryogenesis. EMBO J 20(14):3609–3616. doi:10.1093/emboj/20.14.3609
6. Sabatini S, Beis D, Wolkenfelt H, Murfett J, Guilfoyle T, Malamy J, Benfey P, Leyser O, Bechtold N, Weisbeek P, Scheres B (1999) An auxin-dependent distal organizer of pattern and polarity in the Arabidopsis root. Cell 99(5):463–472
7. Bishopp A, Lehesranta S, Vaten A, Help H, El-Showk S, Scheres B, Helariutta K, Mahonen AP, Sakakibara H, Helariutta Y (2011) Phloem-transported cytokinin regulates polar auxin transport and maintains vascular pattern in the root meristem. Curr Biol 21(11):927–932. doi:10.1016/j.cub.2011.04.049
8. Bishopp A, Help H, El-Showk S, Weijers D, Scheres B, Friml J, Benkova E, Mahonen AP, Helariutta Y (2011) A mutually inhibitory interaction between auxin and cytokinin specifies vascular pattern in roots. Curr Biol 21(11):917–926. doi:10.1016/j.cub.2011.04.017
9. Pacifici E, Polverari L, Sabatini S (2015) Plant hormone cross-talk: the pivot of root growth. J Exp Bot 66(4):1113–1121. doi:10.1093/jxb/eru534
10. Blilou I, Xu J, Wildwater M, Willemsen V, Paponov I, Friml J, Heidstra R, Aida M, Palme K, Scheres B (2005) The PIN auxin efflux facilitator network controls growth and patterning in Arabidopsis roots. Nature 433(7021):39–44. doi:10.1038/nature03184
11. Grieneisen VA, Xu J, Maree AF, Hogeweg P, Scheres B (2007) Auxin transport is sufficient to generate a maximum and gradient guiding root growth. Nature 449(7165):1008–1013. doi:10.1038/nature06215
12. Ibañes M, Fàbregas N, Chory J, Caño-Delgado AI (2009) Brassinosteroid signaling and auxin transport are required to establish the periodic pattern of Arabidopsis shoot vascular bundles. Proc Natl Acad Sci U S A 106(32):13630–13635. doi:10.1073/pnas.0906416106
13. Mattsson J, Sung ZR, Berleth T (1999) Responses of plant vascular systems to auxin transport inhibition. Development 126(13):2979–2991
14. Galweiler L, Guan C, Muller A, Wisman E, Mendgen K, Yephremov A, Palme K (1998) Regulation of polar auxin transport by AtPIN1 in Arabidopsis vascular tissue. Science 282(5397):2226–2230
15. Rubery PH (1977) The specificity of carrier-mediated auxin transport by suspension-cultured crown gall cells. Planta 135(3):275–283. doi:10.1007/BF00384900
16. Rubery PH, Sheldrake AR (1974) Carrier-mediated auxin transport. Planta 118(2):101–121. doi:10.1007/BF00388387
17. Leyser O (1999) Plant hormones: ins and outs of auxin transport. Curr Biol 9(1):R8–R10
18. Fàbregas N, Formosa-Jordan P, Confraria A, Siligato R, Alonso JM, Swarup R, Bennett MJ, Mahonen AP, Caño-Delgado AI, Ibañes M (2015) Auxin influx carriers control vascular patterning and xylem differentiation in Arabidopsis thaliana. PLoS Genet 11(4):e1005183. doi:10.1371/journal.pgen.1005183
19. Sahlin P, Soderberg B, Jönsson H (2009) Regulated transport as a mechanism for pattern generation: capabilities for phyllotaxis and beyond. J Theor Biol 258(1):60–70. doi:10.1016/j.jtbi.2009.01.019
20. Bayer EM, Smith RS, Mandel T, Nakayama N, Sauer M, Prusinkiewicz P, Kuhlemeier C (2009) Integration of transport-based models for phyllotaxis and midvein formation. Genes Dev 23(3):373–384. doi:10.1101/gad.497009
21. Wabnik K, Kleine-Vehn J, Balla J, Sauer M, Naramoto S, Reinohl V, Merks RM, Govaerts W, Friml J (2010) Emergence of tissue polarization from synergy of intracellular and extracellular

auxin signaling. Mol Syst Biol 6:447. doi:10.1038/msb.2010.103

22. van Berkel K, de Boer RJ, Scheres B, ten Tusscher K (2013) Polar auxin transport: models and mechanisms. Development 140(11): 2253–2268. doi:10.1242/dev.079111
23. Abley K, De Reuille PB, Strutt D, Bangham A, Prusinkiewicz P, Maree AF, Grieneisen VA, Coen E (2013) An intracellular partitioning-based framework for tissue cell polarity in plants and animals. Development 140(10): 2061–2074. doi:10.1242/dev.062984
24. Cieslak M, Runions A, Prusinkiewicz P (2015) Auxin-driven patterning with unidirectional fluxes. J Exp Bot 66(16):5083–5102. doi:10.1093/jxb/erv262
25. Rutschow HL, Baskin TI, Kramer EM (2011) Regulation of solute flux through plasmodesmata in the root meristem. Plant Physiol 155(4):1817–1826. doi:10.1104/pp.110.168187
26. Jönsson H, Heisler MG, Shapiro BE, Meyerowitz EM, Mjolsness E (2006) An auxin-driven polarized transport model for phyllotaxis. Proc Natl Acad Sci U S A 103(5):1633–1638. doi:10.1073/pnas.0509839103
27. Smith RS, Guyomarc'h S, Mandel T, Reinhardt D, Kuhlemeier C, Prusinkiewicz P (2006) A plausible model of phyllotaxis. Proc Natl Acad Sci U S A 103(5):1301–1306. doi:10.1073/pnas.0510457103
28. Kramer EM, Frazer NL, Baskin TI (2007) Measurement of diffusion within the cell wall in living roots of Arabidopsis thaliana. J Exp Bot 58(11):3005–3015. doi:10.1093/jxb/erm155
29. Swarup R, Kramer EM, Perry P, Knox K, Leyser HM, Haseloff J, Beemster GT, Bhalerao R, Bennett MJ (2005) Root gravitropism requires lateral root cap and epidermal cells for transport and response to a mobile auxin signal. Nat Cell Biol 7(11):1057–1065. doi:10.1038/ncb1316
30. Goldsmith MH, Goldsmith TH, Martin MH (1981) Mathematical analysis of the chemosmotic polar diffusion of auxin through plant tissues. Proc Natl Acad Sci U S A 78(2): 976–980
31. Kramer EM (2004) PIN and AUX/LAX proteins: their role in auxin accumulation. Trends Plant Sci 9(12):578–582. doi:10.1016/j.tplants.2004.10.010
32. Scarpella E, Marcos D, Friml J, Berleth T (2006) Control of leaf vascular patterning by polar auxin transport. Genes Dev 20(8):1015–1027. doi:10.1101/gad.1402406
33. Liao CY, Smet W, Brunoud G, Yoshida S, Vernoux T, Weijers D (2015) Reporters for sensitive and quantitative measurement of auxin response. Nat Methods 12(3):207–210. doi:10.1038/nmeth.3279, 202 p following 210
34. Bozorg B, Krupinski P, Jönsson H (2014) Stress and strain provide positional and directional cues in development. PLoS Comput Biol 10(1):e1003410. doi:10.1371/journal.pcbi.1003410
35. Kang J, Tang J, Donnelly P, Dengler N (2003) Primary vascular pattern and expression of ATHB-8 in shoots of Arabidopsis. New Phytol 158(3):443–454. doi:10.1046/j.1469-8137.2003.00769.x
36. Press WH, Teukolsky SA, Vetterling WT, Flannery BP (1993) Numerical recipes in FORTRAN; the art of scientific computing. Cambridge University Press, New York
37. Cross M, Greenside H (2009) Pattern formation and dynamics in nonequilibrium systems. Cambridge University Press, New York
38. Fàbregas N, Ibañes M, Caño-Delgado AI (2010) A systems biology approach to dissect the contribution of brassinosteroid and auxin hormones to vascular patterning in the shoot of Arabidopsis thaliana. Plant Signal Behav 5(7):903–906

# Chapter 2

# Strigolactone-mediated Stimulation of Secondary Xylem Proliferation in Stems

Javier Agusti

## Abstract

Secondary xylem (wood) accounts for a large proportion of the terrestrial biomass. Understanding how secondary xylem develops and proliferates is a challenge to enhance our capacities for biomass production. Recent reports revealed that the plant hormone strigolactone is key for the development of secondary xylem. Here, I describe a protocol for strigolactone-mediated stimulation of secondary xylem proliferation in stems. The protocol has been tested in Arabidopsis and Eucalyptus and can be adjusted to other species.

**Key words** Strigolactone, Cambium, Secondary xylem, Lanolin, Stem, Arabidopsis, Eucalyptus

## 1 Introduction

The application of hormonal treatments has been instrumental over decades to enhance biomass production in agriculture. Understanding the precise effect of hormones on specific developmental processes and/or at specific developmental stages allows for establishing new treatments aimed at maximizing biomass yields.

Secondary xylem (wood) is the most abundant biological tissue in the planet. Auxin has long been known to be pivotal in secondary xylem proliferation by initiating the activity of the vascular cambium (*see* **Note 1**). It has been recently discovered that auxin-mediated cambium activity initiation requires strigolactone (SL) signaling and that, indeed, strigolactone alone is sufficient to initiate such cambial activity and to, consequently, stimulate secondary xylem proliferation [1].

Evidences indicate that lack of SL leads to enhanced branching [2] while SL accumulation leads to secondary xylem proliferation [1]. Given that branching and secondary xylem development are two processes resulting in large biomass accumulation, it has been suggested that plants regulate the type of biomass that they generate by regulating the amount of SL that they produce [3]. Therefore, altering the accumulation patterns of strigolactone in specific parts

Miguel de Lucas and J. Peter Etchells (eds.), *Xylem: Methods and Protocols*, Methods in Molecular Biology, vol. 1544,
DOI 10.1007/978-1-4939-6722-3_2, 

of the plant could result in the alteration of biomass accumulation patterns. For example, strigolactone treatments in tree stems could lead to large accumulation of biomass in the form of wood. As a proof of principle of such hypothesis, stems of Arabidopsis plants treated with SL displayed enhanced secondary xylem accumulation [1]. This effect is conserved in trees and, therefore, SL treatments have been suggested as a method to improve our capacities for biomass production in the form of secondary xylem/wood [1]. An exemplar method to induce secondary xylem formation in Arabidopsis and Eucalyptus shoots through SL treatments is described in the next sections. The method can be adapted to other species.

## 2 Materials

### 2.1 Reagents

1. GR24: a strigolactone synthetic analog for which this protocol is optimized. When in powder, GR24 must be kept at room temperature protected from light. GR24 stock solutions must be prepared in acetone (see procedure below) and stored at −20 °C until used to minimize evaporation (*see* **Note 2**).
2. Acetone: used to (1) dissolve GR24 and (2) generate a liquid solution for stem treatments.
3. Tween 20: wet agent for liquid applications of GR24 (*see* **Note 3**).
4. Pen: to mark the treatment zone on the stem.
5. Paintbrush: to perform GR24 liquid applications.
6. Lanolin: a solid carrier for GR24 treatment (*see* **Note 4**).
7. Spatula (or similar tool): to deposit the GR24 on stems when using lanolin as a carrier.
8. Weighing boats: to (1) weight GR24 on a precision balance and (2) use as a support for mixing lanolin with GR24.
9. Soft tissue: to remove lanolin from the stems.
10. Razor blade: to collect treated stems for further analyses.
11. Paraformaldehyde (PFA): to generate a fixative for collected samples (for further microscopy analyses).
12. Microtome: to section samples for analyses.
13. Slides: to mount sectioned samples.
14. Toluidine blue: to stain sectioned samples.
15. Coverslips: to cover stained samples on slides.
16. Microscope: to observe/analyze samples.

### 2.2 Solutions

#### 2.2.1 Preparing GR24 Stock Solution

1. Using a precision balance, weight sufficient GR24 to generate a stock solution with a final concentration between 5 and 10 mM.
2. Rapidly, solve the GR24 in the required volume of acetone to reach the desired concentration (at room temperature) and store the solution immediately at −20 °C.

#### 2.2.2 Preparing GR24 Liquid Solution

1. Dilute GR24 from stock solution to the desired concentration (typically between 1–10 μM; *see* **Note 5**) in a 0.5 % acetone and 0.1 % Tween 20 solution.
2. A 0.5 % acetone and 0.1 % Tween 20 solution should be used as mock control.

### 2.3 Preparing GR24 Treatments Using Lanolin as Solid Carrier

1. *GR24 in lanolin*:
   (a) Heat lanolin (usually contained in a glass bottle) in a microwave until melted to liquid state.
   (b) Pipette 1 ml and transfer to a weighing boat.
   (c) Let lanolin cool down.
   (d) Before lanolin completely solidifies, add the appropriate volume of GR24 from the stock solution to reach the desired concentration (typically between 1–10 μM).
   (e) Mix well and proceed immediately to treat plants.
2. *Mock*:
   (a) Heat lanolin (as described in **step 1(a)**).
   (b) Pipette 1 ml and transfer to a weighing boat.
   (c) Let lanolin cool down.
   (d) Before lanolin completely solidifies, add the appropriate amount of acetone (exactly the same volume as the GR24 stock solution volume used in **step 1(d)**).

## 3 Method

All procedures should be carried out at room temperature unless otherwise stated.

### 3.1 Plant Material and Growth Conditions

All plants should grow in long day conditions (16 h light, 8 h dark) at 21 °C and with standard light. For both *Arabidopsis* and *Eucalyptus*, plants were treated when they reached a stature of 5–10 cm (*see* **Note 6**).

### 3.2 Strigolactone Treatments

Perform treatments on stems as described below. Treatments should be repeated every second day during 21 days (*see* **Note 7**), always in (exactly) the same position within the stem (Fig. 1). In

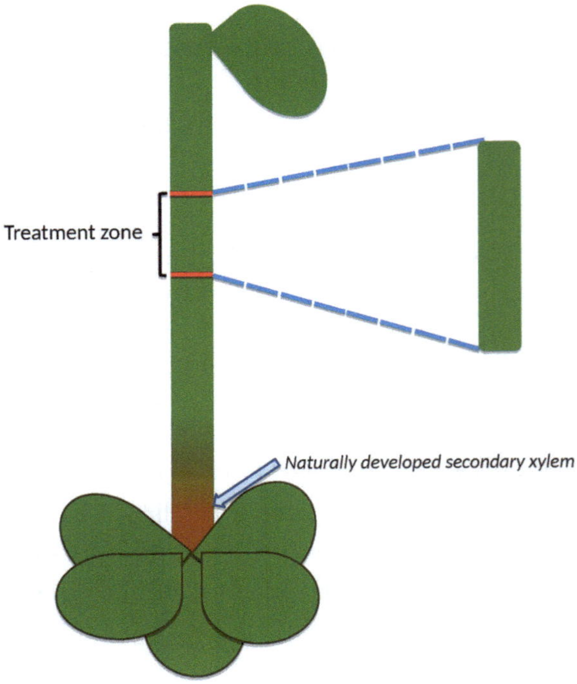


**Fig. 1** Strigolactone treatment on Arabidopsis stem. *Red stripes* mark the treatment zone within the stem. The zone where secondary xylem develops in a natural manner is marked in *brown*. In Arabidopsis stems such zone comprises typically ~1 cm from the base of the stem in adult plants. Treated zones must be collected using a razor blade for further analyses

order to ensure that the treated zone in the stem is always exactly the same, it is recommendable to previously mark such zone using a regular marker. The treatment zone must be located in a position where secondary xylem does not develop in a natural manner (Fig. 1). GR24 can be applied in a liquid solution or using lanolin as a carrier. Both types of treatments work equally well in *Arabidopsis* and *Eucalyptus*. However, if working with other species it is recommendable to determine whether it is more convenient to apply GR24 in a liquid solution or with lanolin (depending on the nature of the species and on its stem morphology).

*Treatments with GR24 in Liquid Solution*

1. Determine the treatment position within the stem and mark it with the help of a pen (as described in Fig. 1).
2. Generate new, fresh liquid treatment solutions (0.5 % acetone and 0.1 % Tween 20 in water) before each treatment.
3. Separate an aliquot for mock control and an aliquot for GR24 treatment. Add the required amount of GR24 (from the stock solution) to the latter to reach the desired concentration (*see* **Note 5**).

4. Mix well and proceed immediately to treat the predetermined treatment zone of the stem (see **step 1** of this subheading) by literally "painting" it with the liquid solution, with the help of a paintbrush.
5. Perform mock control treatments in the same way (using the mock control solution). To avoid GR24 contaminations in the mock treatments, always use a different paintbrush to the one used for GR24 treatments.
6. Repeat GR24 and mock control treatments every second day, always in the exact same position of the stem, for 21 days.

*Treatments with GR24 in Lanolin*

1. Determine the treatment zone within the stem, as described in Fig. 1.
2. Apply the GR24 mixed in lanolin (generated as described in Subheading 2.2) on the treatment zone using a spatula.
3. Perform mock treatments exactly in the same manner.
4. Repeat GR24 and mock control treatments every second day, always in the exact same position of the stem, for 21 days. Before repeating each treatment, lanolin from the previous treatment must be gently removed using soft tissue.

### 3.3 Tissue Harvesting and Histological Analyses

Before harvesting, prepare a fresh aqueous solution of 4% paraformaldehyde (PFA) to be used as fixative (*see* **Note 8**).

1. Collect the treated area of the GR24 and mock treated plants with a razor blade (Fig. 1) and transfer the samples immediately to the fixative.
2. Fix samples in 4% PFA by (1) vacuum infiltration (15 min) and (2) overnight incubation at 4 °C.
3. Store samples at 4 °C until used (still within the fixative solution).
4. Process samples for wax embedding as described [4, 5].
5. Section samples using a microtome, mount them on water on slides, and leave at 42 °C overnight.
6. For best observation, staining with toluidine blue (0.05%) is recommended, following previously described protocols [4, 5].
7. Cover slides with coverslips and mounting medium and observe slides in a microscope using bright field and white light. Compare the anatomy of the mock and GR24 treated samples.

## 4 Notes

1. The vascular cambium is the meristem that gives rise to the secondary vascular tissues (i.e., secondary xylem and secondary phloem).
2. Acetone evaporates very quickly at room temperature; this is the reason why stock solutions should remain at −20 °C (acetone remains liquid at this temperature). However, even at −20 °C acetone experiences some evaporation. This is a problem because it can result in changes in the actual concentration of the stock solution. To solve this problem, it is recommended to check regularly the volume of the stock solution and to keep a strict track of the amount of stock solution used. Replace acetone if needed to keep the right concentration.
3. For this protocol we used Tween 20 as wet agent. However, other wet agents can be used without altering the effect.
4. Lanolin is an inert form of fat that has been extensively used as solid carrier for local application of specific chemicals and/or hormones in specific organs or spatial locations in plants. Lanolin alone has no effect on plants.
5. A GR24 1–10 μM concentrations is the best concentration range for xylem proliferation in Arabidopsis and Eucalyptus. However, the optimal concentration to be used might be different in other species.
6. GR24 treatment tends to work better in young plants. In Arabidopsis and Eucalyptus, the treatment yielded best results when applied on plants that were 5–10 cm tall.
7. Shorter treatments did not induce cambium activity initiation or xylem proliferation. It has been suggested that GR24 may be difficult to be taken up by epidermal cells in the stem. Therefore, treatments need to be prolonged in time.
8. Other fixatives (i.e., 70% EtOH) can be used.

### References

1. Agusti J, Herold S, Schwarz M, Sanchez P, Ljung K, Dun EA, Brewer PB, Beveridge CA, Sieberer T, Sehr EM, Greb T (2011) Strigolactone signaling is required for auxin-dependent stimulation of secondary growth in plants. Proc Natl Acad Sci U S A 108(50):20242–20247
2. Gomez-Roldan V, Fermas S, Brewer PB, Puech-Pages V, Dun EA, Pillot JP, Letisse F, Matusova R, Danoun S, Portais J-C, Bowmeester H, Becard G, Beveridge CA, Rameau C, Rochange SF (2008) Strigolactone inhibition of shoot branching. Nature 455:189–195
3. Agusti J, Greb T (2013) Going with the wind—adaptive dynamics of secondary meristems in plants. Mech Dev 130(1):34–44
4. Agusti J, Lichtenberger R, Schwarz M, Nehlin L, Greb T (2011) Characterization of transcriptome remodelling during cambium formation identifies *MOL1* and *RUL1* as opposing regulators of secondary growth. PLoS Genet 7(2):e1001312. doi:10.1371/journal.pgen.1001312
5. Sehr EM, Agusti J, Lehner R, Farmer EE, Schwarz M, Greb T (2010) Analysis of secondary growth in the Arabidopsis shoot reveals a positive role of jasmonate signaling in cambium formation. Plant J 63:811–822

# Chapter 3

# Quick Histochemical Staining Methods to Detect Cell Death in Xylem Elements of Plant Tissues

**Sacha Escamez, Benjamin Bollhöner, and Hannele Tuominen**

## Abstract

Histochemical assays of xylem cell death cannot take advantage of the conventional methods for detection of cell death, such as staining with propidium iodide or trypan/Evans blue or the TUNEL staining. This chapter presents two alternative histochemical methods that can be used to detect xylem cell death quickly and reliably using light microscopy. The first method is a viability stain that can be used to detect cell death of different types of xylem elements in basically any plant species. The second method reveals cell death in xylem vessel elements based on their functionality in transport of water and small water-soluble stains.

**Key words** Xylem, Wood, Programmed cell death, Tracheary element, Libriform fiber, *Populus*, Arabidopsis, Viability stain, Sulforhodamine, Nitroblue tetrazolium

## 1 Introduction

Cell death is a critical stage of xylem differentiation. It is important to have good methods to detect the occurrence of cell death not only for the purpose of increasing the basic understanding of xylem differentiation but also since cell death has implications on growth and the properties of the xylem. The most reliable way to detect the progression of cell death in xylem tissues is to use genetic markers. Fluorescent markers constructed either by transcriptional or translational fusions to sequences derived from genes like the aspartic protease *PASPA3*, *BIFUNCTIONAL NUCLEASE 1/ENDO1*, or *METACASPASE9* [1, 2] are indicative of the different stages of xylem maturation and cell death. Using genetic markers is however time consuming and possible only in a few species. Histochemical methods are a good alternative to the genetic methods. Histochemical detection of cell death depends on either (1) identification of cells that are living in contrast to dead cells or (2) identification of cells that are dead or dying in the midst of tissues that are otherwise composed of viable cells. This chapter presents

Miguel de Lucas and J. Peter Etchells (eds.), *Xylem: Methods and Protocols*, Methods in Molecular Biology, vol. 1544,
DOI 10.1007/978-1-4939-6722-3_3, © Springer Science+Business Media LLC 2017

two methods that can be used to detect cell death based on these two criteria in the xylem tissues of plants.

Histochemistry of xylem tissues is challenging for many reasons. The histochemical methods that are commonly used to detect cell death in animal systems are dependent on good penetration of substrates or stains for cell death related enzymes, molecules, or physiological conditions. Xylem tissues are usually located furthest away from the surface of the plant organs, which hinders penetration of stains commonly used to detect apoptotic cell death such as fluorescein diacetate (FDA) or the DNA binding propidium iodide (PI). These are in the animal systems often used in combination to detect simultaneously viability of the cells by FDA hydrolysis in living cells and cell death by the loss of plasma membrane integrity and concomitant binding of PI to DNA [3]. PI can be used to detect loss of plasma membrane integrity in the xylem elements of the young part of plant roots before formation of the casparian strip which prevents its further penetration into the xylem elements. We have however only rarely seen PI staining of nuclei in xylem elements of the non-suberized part of the root, which is most probably due to the fact that the nucleus is degraded within 20 min after bursting of the vacuole at the final stages of xylem maturation [4]. It is simply difficult to "catch the moment," making PI unsuitable for in situ detection of plasma membrane integrity and cell death in xylem elements. Combination of FDA and PI, however, works perfectly well for cell cultures like the xylogenic cultures that can be induced to differentiate into tracheary elements (Fig. 1) [5]. Even though we rarely see PI-positive nuclei in these cell cultures, dead cells can be conveniently detected with PI since it also binds to the cell walls and the cellular remnants of the dead cells (Fig. 1; *see* also Ref. 5). Another traditional method for detection of cell death in both animals and plants is exclusion of dyes such as trypan blue or Evans blue by an intact plasma membrane [6]. This type of staining is not suitable for xylem tissues either since these dyes do not have good penetration properties. The same is true for the very popular assay for detection of cell death, the Terminal deoxynucleotidyl transferase dUTP nick end labeling (TUNEL), which detects breaks in the DNA. We have tried to improve the penetration of the substrates and the terminal deoxynucleotidyl transferase enzyme into the root stele of Arabidopsis seedlings without success.

In order to circumvent the problems with tissue penetration of the different stains into the xylem elements, histochemical staining can be performed on tissue sections. In this case, it is crucial to take into consideration the great length of the xylem elements and the fact that the sectioning breaks all xylem elements that are exposed to the surface of a section. A solution to this problem can

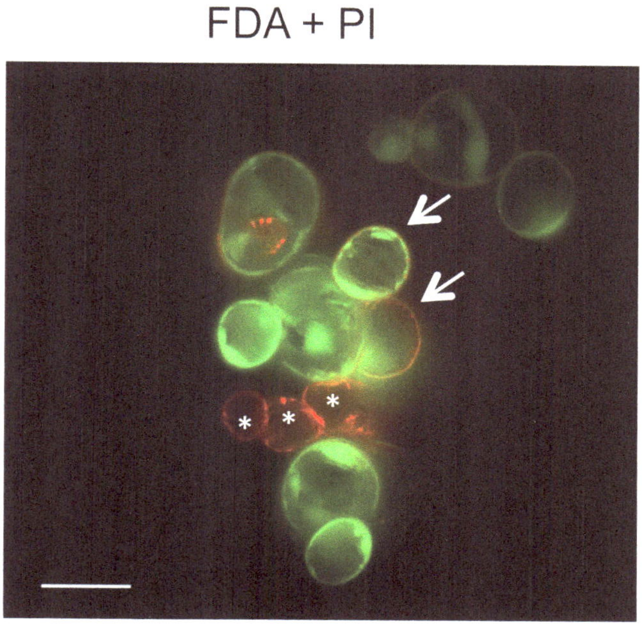


**Fig. 1** Combined staining with fluorescein diacetate (FDA) and propidium iodide (PI) in xylogenic Arabidopsis cell culture 5 days after hormonal induction. The green fluorescence from fluorescein that is produced by hydrolysis of FDA in living cells reveals that most of the cells of the culture are still alive. PI penetrates the cells that have compromised plasma membrane integrity and also binds to the remnants of dead cells. The *asterisks* indicate dead cells. The *arrows* indicate cells that are either very close to the moment of cell death or that have just died, as judged from the advanced stage of the secondary cell wall patterning and PI staining. Scale bar = 50 μm

be longitudinal sections which, however, many times are difficult to interpret due to problems in section orientation and tissue overlays (*see* Fig. 3f–h in Ref. 7). Many of these problems are circumvented by using stains that detect compounds like RNA or enzyme activities throughout the entire length of the xylem elements. For RNA, a commonly used dye is acridine orange [8] but which also binds strongly to the secondary cell wall of xylem elements. We present here a technique (Procedure 3.1) that has been used with plant tissues already in the 1970s [9, 10] but that has maybe not reached a wide audience. It is based on detection of succinate dehydrogenase activities of living cells [11]. These activities are most abundant in the mitochondria of living cells. The advantages of this technique are that (1) it is simple and fast, (2) it does not rely on fluorescence detection, avoiding the problems linked to autofluorescence of lignin and other fluorescing compounds inherent to xylem, and (3) it normally results in strong staining in all types of living tissues. We also present assays for detection of dead vessel elements (Procedure 3.2) on the basis of their functionality in transport of water and water-soluble compounds. Histochemical staining techniques for the xylem vessel transport of two compounds, Evans blue and sulforhodamine B, are presented in this chapter. Both these stains are water soluble and flow easily along with the vessel transpiration stream, but sulforhodamine B binds transiently to the cell wall making it more suitable for detection of vessel functionality in histochemical assays. Both procedures are suitable for both herbaceous and woody species [7, 12, 13].

## 2 Materials

### 2.1 Detection of Xylem Cell Viability

1. Reaction buffer: Mix equal amounts of 0.2 M sodium succinate and 0.2 M phosphate buffer, pH 7.6. For a 10 ml total volume of the reaction buffer, add 270 mg sodium disuccinate dibasic hexahydrate in 5 ml water, and mix with 5 ml of 0.2 M phosphate buffer, pH 7.6. The buffer can be stored up to 3 months at 4 °C.
2. Reaction substrate: Mix 10 mg nitrotetrazolium blue chloride (NBT) in 10 ml water prior to the experimental Procedure.

### 2.2 Detection of Vessel Cell Death by Their Function in Water Transport

1. Evans blue staining solution: 0.05–0.25% Evans blue in water or Murashige and Skoog medium.
2. Sulforhodamine B staining solution: 0.01% sulforhodamine B in Murashige and Skoog medium.

## 3 Methods

### 3.1 Detection of Xylem Cell Viability

1. Prepare the test solution by adding one part of reaction buffer with one part of the freshly prepared reaction substrate (*see* **Note 1**).
2. Prepare transverse sections or radial sections from the vascular tissues of plants (*see* **Note 2**) by hand sectioning (Fig. 2) (*see* **Note 3**). The thickness of the section should be optimized for each plant material individually (*see* **Note 4**).
3. Place the sections in flat-bottom multiwell plates or glass vials filled with the test solution, swift gently and cover with a lid. Incubate in good light conditions (*see* **Note 5**) at room temperature or 37 °C as long as an intensive blue staining is visible in the sections (*see* **Note 6**).
4. Stop the reaction by rinsing the sections carefully with water. Pick up the sections on an object slide, mount in water or in 50% glycerol (*see* **Note 7**).

**Fig. 2** Hand sectioning with the help of a sample holder. A piece of a *Populus* stem is fixed into the sample holder with the help of two metal clamps

5. Document the results immediatedly with a light microscope. Viable cells can be identified by their blue staining (Fig. 3; *see* **Note 8**).

#### 3.2 Detection of Vessel Cell Death by Their Function in Water Transport

1. Sampling of woody species: Collect the apical part of the shoot by cutting with a sharp knife under water and place the shoot immediately in the staining solution (*see* **Note 9**).
2. Sampling of Arabidopsis inflorescence stems: Remove flowering plants from the soil, cut off the root 1 cm under the rosette under water and place the plant on top of a low petri dish filled with the staining solution so that the leaves are supported by the edges of the petri dish and the hypocotyl is immersed into the staining solution.
3. Incubate in the staining solution at room temperature in conditions that are favourable for water transpiration, for 1 h or until the stain has reached the area of interest (*see* **Note 10**).
4. Make transverse sections by hand (*see* **Note 11**) from the area or interest (*see* **Note 8**) and place them into object slides. Evans blue stained sections should be mounted without mounting medium or in a very little volume of water or 50% glycerol (*see* **Note 12**). Sulforhodamine B stained sections can be mounted in water or 50% glycerol (*see* **Note 13**).
5. Examine immediately under light microscope. Functional vessels appear blue with Evans blue and bright pink with sulforhodamine B (Fig. 4; *see* **Note 14**).

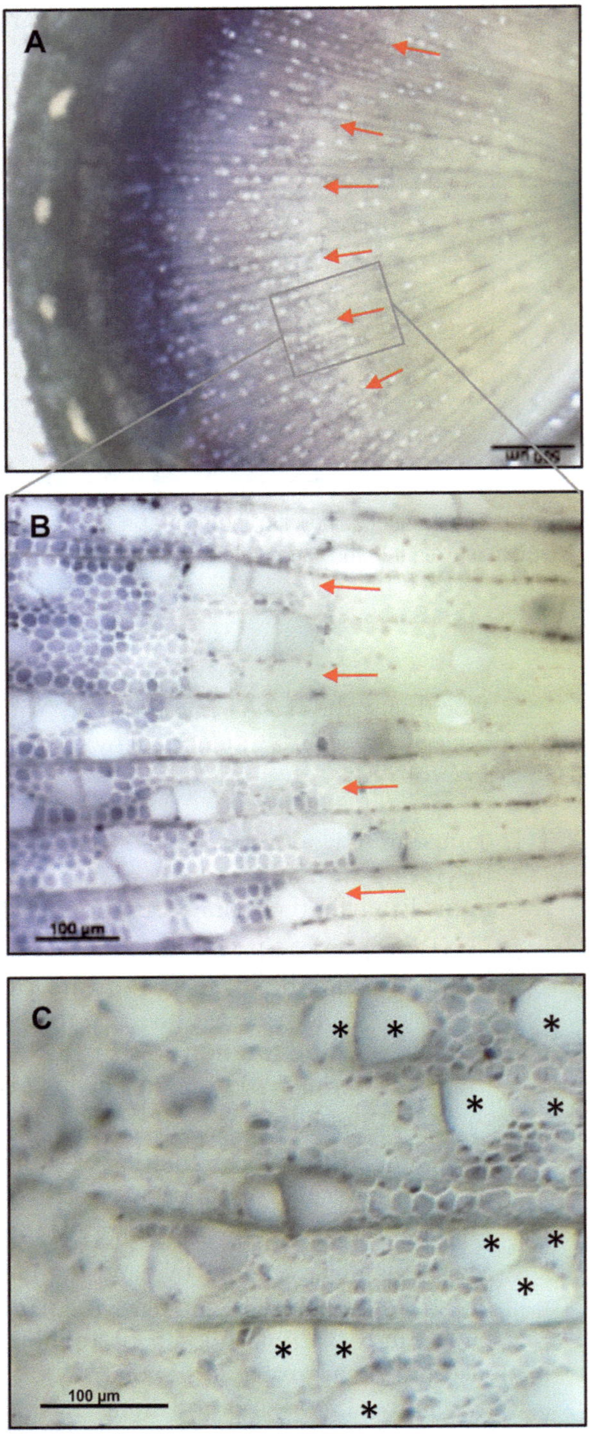


**Fig. 3** Histochemical staining of living vessels and fibers with nitroblue tetrazolium (NBT) in transverse sections of *Populus* stem. (**a**–**c**). Images are shown for transverse sections from the basal part of the stem of 2-month-old *Populus* trees. (**a**) Xylem fibers die in a rather synchronous manner around the circumference of the stem at a distance of around 1 mm from the vascular cambium. (**b**) is a magnification of the area indicated with a square in (**a**). (**c**) Vessel elements die at a distance of 300–500 μm from the cambium (which is outside of the image to the left). The *red arrows* point to the zone in the wood where most of the fibers die. The *asterisks* indicate dead vessel elements. Scale bars 500 μm (**a**) and 100 μm (**b**, **c**)

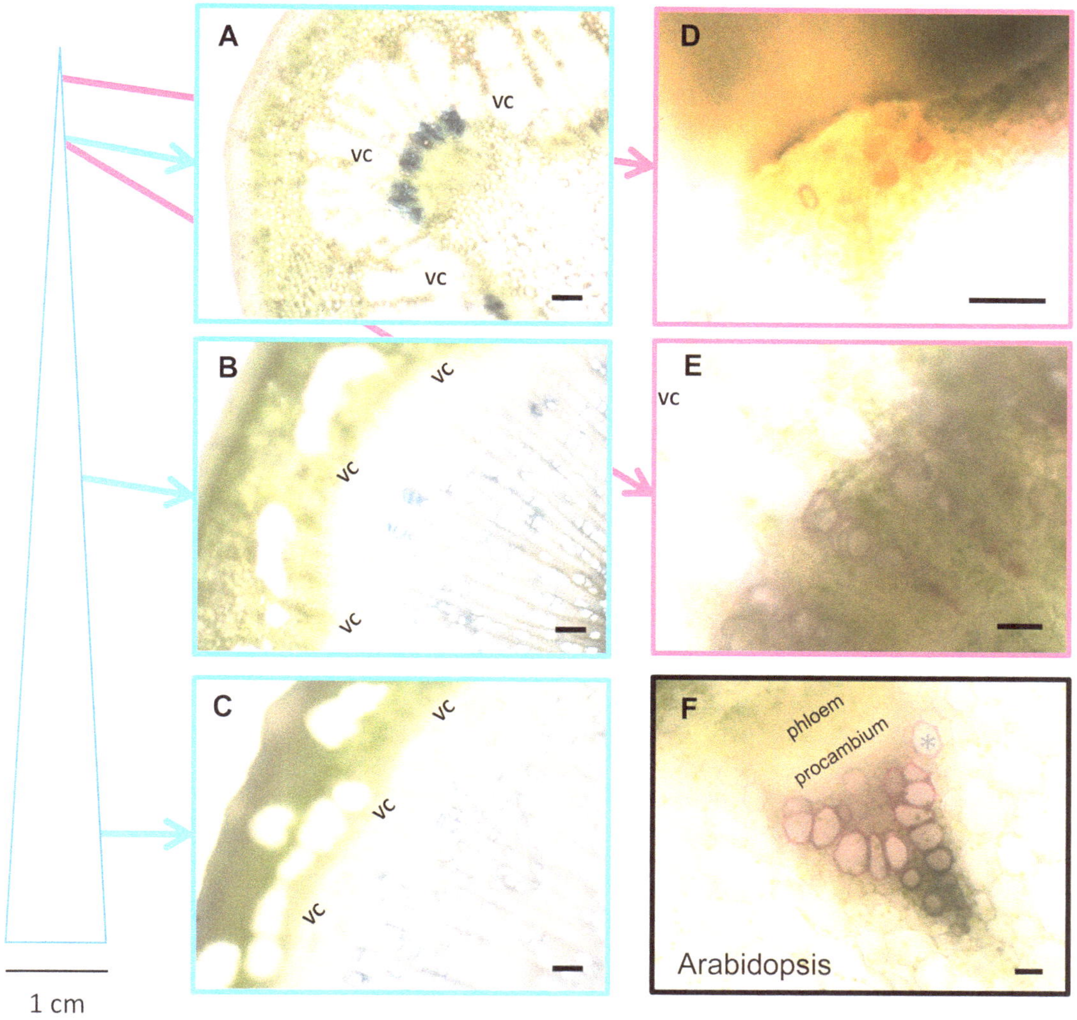


**Fig. 4** Histochemical staining of dead vessels on the basis of their functionality. Images are shown for transverse sections of *Populus* shoots that were incubated in Evans blue (**a**–**c**) or sulforhodamine B (**d**, **e**), and for an Arabidopsis inflorescence stem that was derived from plants incubated for 1 h in sulforhodamine B. The *Populus* shoots were 10-cm-long for the sulforhodamine staining and 50-cm-long for the Evans blue staining. The transverse sections from the *Populus* shoots were taken as close to the apex as possible (**d**), from the seventh internode (**a**, **e**), from the middle part (**b**) and from the base (**c**) of the shoot. The staining in (**b**) and (**c**) reveals that the vessel elements die and start transporting water within the distance of 300 μm from the cambium in this case. The staining in (**d**) reveals that functional vessel elements are present in the first transverse sections that can be taken by hand in the apical part of the *Populus* stem. The *asterisk* in (**f**) indicates a recently functionalized vessel element. *VC* vascular cambium. Scale bars 100 μm (**a**, **b**, **c**, **f**), 50 μm (**d**), 20 μm (**e**)

## 4 Notes

1. An alternative substrate is 2,3,5-triphenyltetrazolium chloride which has also been reported suitable for quantification of the signal [14].
2. Remember that the sectioning normally breaks all the xylem elements since the xylem elements are typically longer than the average section thickness. Handle the sections gently to minimize the risk of flushing away the cell contents. An alternative is to stain larger tissue pieces and make sections only after staining of the whole tissue piece, which allows analysis of intact xylem elements. In order to secure the intact nature of all xylem elements, the sectioning of the stained tissue pieces should be done at a distance from the surface of the tissue piece that overrides the length of the xylem elements. This method gives more reliable results.
3. Vibratome sectioning gives generally a better and a more even quality of the sections compared to hand sectioning, but does not in our hands result in thick enough sections for the NBT staining.
4. A suitable thickness to start with is around 50 μm. Thinner transverse sections tend to lose the cellular contents, and thicker sections become anatomically difficult to analyze.
5. Even though the original protocol [11] suggests incubation in darkness, we experience faster accumulation of the reaction product when incubated in good light conditions, for instance under a table lamp.
6. The time of incubation should be optimized again for each sample type separately. A good typical time to start with is 1 h. Too long incubation leads to product precipitation in the buffer and unspecific binding to the samples.
7. The slides are not permanent. It is reported that permanent slides can be made by incubating the samples in Zirkle's fixative (Mix 25 mg potassium dichromate, 25 mg ammonium dichromate and 20 mg copper(II) sulfate in a total of 4 ml water) for 24–48 h followed by dehydration and mounting in euparal.
8. The life time of the xylem elements can be estimated by measuring the number of the living cells in a radial file of xylem elements (Fig. 3a and c) and dividing that by the rate of cell division. Different genotypes (with a similar cambial division rate) can be compared to each other by measuring the distance from the procambium or cambium to the position where most xylem elements die. In order to do this, it is crucial that the cambial zone and the xylem expansion zone are not crushed or squashed by sectioning.

9. If very long (50–70 cm) shoots of woody species are stained, prevent drying of the leaves by increasing the humidity of the air.
10. For Arabidopsis inflorescence stems and young parts of woody species, normally 1-h incubation time is enough for the stain to reach the first functional vessel elements in the apex (see Fig. 4d and f).
11. Evans blue stained sections should not be too thin not to lose the stain. Sulforhodamine B stained sections can be as thin as possible.
12. Evans blue is water-soluble and is easily flushed away from the sections if there is too much water.
13. Even though sulforhodamine B is water soluble, it is transiently attached to the cell walls and therefore not as easily removed from the vessel elements as Evans blue.
14. Sulforhodamine B is a fluorescent dye (maximal absorbance at 565 nm and maximal emission at 585 nm) and can be detected with epifluorescence or confocal microscope. In our view, however, this is unnecessary due to its distinct color in bright light.

## Acknowledgment

This work was supported by grants 232-2009-1698 and 232-2011-1312 from the Swedish Research Council Formas, and grant 621-2013-4949 from the Swedish Research Council VR. We wish to acknowledge the Plant Cell Wall Laboratory at Umeå Plant Science Centre.

### References

1. Fendrych M, Van Hautegem T, Van Durme M, Olvera-Carrillo Y, Huysmans M, Karimi M, Lippens S, Guérin CJ, Krebs M, Schumacher K, Nowack MK (2014) Programmed cell death controlled by ANAC033/SOMBRERO determines root cap organ size in Arabidopsis. Curr Biol 24:931–940
2. Bollhöner B, Zhang B, Stael S, Denancé N, Overmyer K, Goffner D, Van Breusegem F, Tuominen H (2013) Post mortem function of AtMC9 in xylem vessel elements. New Phytol 200:498–510
3. Jones KH, Senft JA (1985) An improved method to determine cell viability by simultaneous staining with fluorescein diacetate-propidium iodide. J Histochem Cytochem 33:77–79
4. Obara K, Kuriyama H, Fukuda H (2001) Direct evidence of active and rapid nuclear degradation triggered by vacuole rupture during programmed cell death in Zinnia. Plant Physiol 125:615–626
5. Escamez S, André D, Zhang B, Bollhöner B, Pesquet E, Tuominen H (2016) METACASPASE9 modulates autophagy to confine cell death to the target cells during Arabidopsis vascular xylem differentiation. *Biol Open* 2016 Jan 6 [Epub ahead of print]. doi:10.1242/bio.015529
6. Greenberg JT, Ausubel FM (1993) Arabidopsis mutants compromised for the control of cellular damage during pathogenesis and aging. Plant J 4:327–341
7. Courtois-Moreau CL, Pesquet E, Sjödin A, Muñiz L, Bollhöner B, Kaneda M, Samuels L, Jansson S, Tuominen H (2009) A unique program for cell death in xylem fibers of Populus stem. Plant J 58:260–274
8. Stadelmann EJ, Kinzel H (1972) Vital staining of plant cells. In: Prescott DM (ed) Methods in cell physiology, vol 5. Academic, New York, pp 325–372

9. Berlyn GP, Miksche JP (1976) Botanical microtechnique and cytochemistry. Iowa State University Press, Ames, IA
10. Gahan PB (1984) Plant histochemistry and cytochemistry: an introduction. Academic, London
11. Fine IH, Costello LA (1963) The use of starch electrophoresis in dehydrogenase studies. Methods Enzymol 6:958–972
12. Turner SR, Hall M (2000) The gapped xylem mutant identifies a common regulatory step in secondary cell wall deposition. Plant J 24: 477–488
13. Zhao C, Avci U, Grant EH, Haigler CH, Beers EP (2008) XND1, a member of the NAC domain family in Arabidopsis thaliana, negatively regulates lignocellulose synthesis and programmed cell death in xylem. Plant J 53:425–436
14. Towill LE, Mazur P (1975) Studies on reduction of 2,3,5-triphenyltetrazolium chloride as a viability assay for plant-tissue cultures. Can J Bot 53:1097–1102

# Chapter 4

# Establishment and Utilization of Habituated Cell Suspension Cultures for Hormone-Inducible Xylogenesis

**Delphine Ménard, Henrik Serk, Raphaël Decou, and Edouard Pesquet**

## Abstract

The development of inducible cell differentiation in suspension cultures led to multiple breakthroughs. It enabled the understanding of the chronology, duration, regulation and interdependency of the multiple events leading to fully functional specialized cells. The most studied cell differentiation in plants using inducible suspension cultures is the formation of tracheary elements (TEs) - the hydro-mineral sap conducting cells. Several in vitro systems established in different plant species have been developed to trigger TE formation on-demand. Here, we describe the establishment, harvesting and analysis of *Arabidopsis thaliana* stable habituated cell lines inducible by hormones to differentiate into TEs on-demand. Moreover, we explain the means to monitor and modify the chronology, duration and regulation of the progression of TE formation.

**Key words** Tracheary elements, Habituated cell cultures, Hormone inducible differentiation, *Arabidopsis thaliana*

## 1 Introduction

Understanding the chronology and duration of the differentiation events occuring in specific cell types within pluricellular organisms has always been the goal of plant cell and developmental biology. Yet, plant indeterminate growth, the sub-epidermal localization of cells and the cell-to-cell intra/interdependency hinder the synchronized and non-destructive monitoring of the differentiation sequence of specific cell types. This is particularly true when studying the formation of the xylem hydro-mineral sap conducting cells in higher vascular plants, called tracheary elements (TEs) [1–4]. These specific cells conduct the hydro-mineral sap by undergoing a well-defined differentiation program which requires that: (1) the side walls of the TE cell are reinforced with the deposition of a circumferential patterned secondary cell wall [1–4], (2) the TE content is hollowed out by undergoing programmed cell death [1–5] and (3) the TE terminal ends are thinned or perforated to

Miguel de Lucas and J. Peter Etchells (eds.), *Xylem: Methods and Protocols*, Methods in Molecular Biology, vol. 1544, DOI 10.1007/978-1-4939-6722-3_4, interconnect lumens to form a continuous vascular cylinder [1–4]. The hydro-mineral sap thus flows transcellularly through the hollowed TE corpses which are laterally delimited by patterned secondary reinforcements maintaining the conduit at a constant diameter, independently of the vascular sap pressure [4]. The main difficulties faced when studying TE development at the cellular level in whole plants are that: (1) xylem is an internal tissue experimentally accessible only by destructive means, (2) xylem functions by using dead TE cells thus only few elements are actively undergoing differentiation and even less are at the same developmental state of TE formation, and (3) inhibition of TE development in whole plants can result in collapsed TEs unable to withstand the negative pressure associated to the hydro-mineral sap flow. As altering TE formation leads to an overall disturbance of the plant physiology and development, the study of TE formation in whole plants is difficult when using classical experimentation methods.

The development of inducible in vitro systems from plant cell cultures to form TEs on-demand has therefore provided an alternative bypassing these experimental difficulties. The capacity to induce TE differentiation derived from the initial studies of Vöchting [6] who showed that pith parenchyma cells differentiate into TEs to connect the vascular system of scion and stock during grafting. Since the mid-1960s, simplified systems from in vitro cultures from multiple plant species have been developed to induce TE differentiation using either explants [7, 8], calli [9–12] or cell suspension cultures [13–20]. Independently from the species used, the differences between these in vitro systems comprise (1) the proportion of initial cells differentiating into TEs—ranging from a couple of percent for the *Lactuca* explants system [8] to 50–60% in the habituated *Arabidopsis* system [18], (2) the timing to reach the TE differentiation maximum—ranging from less than a week for the *Zinnia* system [13] to several weeks for the *Cryptomeria* calli system [12], (3) the synchrony with which TE cell differentiation reaches the plateau—ranging from continuous [7, 8] to biphasic [13] or monophasic progressions [18], (4) the individualization of TEs from cell clumps—ranging from TEs included within cell clumps/calli in *Arabidopsis* stable lines [7, 8, 14, 15] to single isolated cells in the *Zinnia* or *Arabidopsis* systems [13, 18], (5) the genetic transformation tools—either through transient transformation by electroporation for *Zinnia elegans* cells [21] or stable transformation by coculture with *Agrobacterium* for *Arabidopsis* cell lines [15, 16, 18], and (6) the amount of biological material produced by the system during the progression of TE differentiation—ranging from 10–100 mg for the *Zinnia* system [13] to 100 g for the habituated *Arabidopsis* cell suspensions [18]. The most used and performant biological systems are derived from cell suspension cultures which can be divided into two groups depending on the means to induce TEs. On one hand, systems inducible by hormones which include (1) freshly isolated mesophyl cells of *Zinnia elegans* (with auxin and cytokinin) [13], (2) *Arabidopsis thaliana* 2,4D-maintained cell lines (washed and supplied with brassinolides and boric acid) [14] and (3) *Arabidopsis thaliana* habituated cell suspensions (with auxin, cytokinin, and brassinolides) [18]. On the other hand, systems inducible by chemicals triggering the overexpression of TE master transcription regulator genes such as the dexamethasone-inducible VND7 in tobacco BY-2 cells [20] or estradiol-inducible VND6 in *Arabidopsis thaliana* cells [16].

Far from comparing all the different TE differentiation systems, we herein present the experimental procedures to establish and use *Arabidopsis thaliana* habituated cell suspension cultures capable of TE differentiation on-demand as initially established by Pesquet in 2010 [18]. Beyond the fact that *Arabidopsis thaliana* is the genetic plant model, this biological system fulfills many optimal features including (1) a high differentiation efficiency ranging from 30–50% of all cells [18, 22], (2) a quick differentiation process achieved in less than 1 week [18, 22], (3) a high level of synchrony with a monophasic progression to reach the differentiation plateau [18, 22], (4) a single cell system in which individualized TEs can be directly monitored [18, 22, 23], (5) stable genetic transformation tools for promoter analysis, constitutive silencing and over-expression of fusion proteins [18, 22, 23], and (6) large amounts of cells produced with ease as the system derives from stable primary cell suspensions cultures [22]. This system has been successfully used to make breakthroughs in understanding TE formation. Its combined use with long-term live cell imaging enabled to unravel the relationship between cell death and the deposition of secondary wall reinforcements at the single cell level [18, 22–24]. This system also allowed high through-put analysis, such as transcriptomics and quantitative proteomics, during the progression of the synchronized differentiation into TE cells [22]. Finally, this system was used for reverse genetic analysis of gene function using constitutive RNA silencing during TE differentiation [18, 22–24]. We herein describe detailed step-by-step methods and possible pit falls associated with the establishment and uses of *Arabidopsis thaliana* habituated cell suspension cultures for TE differentiation on demand.

## 2 Materials

All the chemicals were mostly purchased from Sigma (other companies will be specified otherwise). Conscientiously, all chemical and glassware wastes as well as GMO were decontaminated and disposed using the appropriate procedure. A routine procedure to minimize cell culture wastes is to incubate cells at room temperature without agitation overnight, then filter the cell suspension through cellulose paper, autoclaving the flow-through culture medium and placing the filtered cell solid waste in GMO containers for decontamination.

### 2.1 Equipment and Infrastructure

1. Sterilization and sterile manipulation:
   All manipulations of plant explants, calli and cell suspensions need to be performed in a sterile environment using a sterile laminar flow hood such as a ScanLaf class 2 Mars Premium 1200 (Labogene, Denmark). All polypropylene plastics (bottle caps, microtubes, pipette tips...) and open glassware need to be covered by a double layer of aluminum foil or enclosed in autoclavable containers and sterilized by autoclaving (such as using a Tuttnauer 3870EL horizontal autoclave). Clean and dry cell culture (Erlenmeyer) flasks are closed with a cotton wool ball in the flask mouth and covered up to the neck with a double layer of aluminum foil prior to autoclaving. Scalpels with blades (no. 22 surgical blade) and tweezers (Bochem stainless steel 10/18) are first wrapped in a double layer of aluminum foil and sterilized by autoclaving prior to use. During their use under the laminar flow, scalpel blades and tweezers are flame-sterilized using a Bunsen burner (Fireboy Portable Safety Bunsen Burner, Integra, Switzerland) in-between samples and left to cool on a surface sterilized metal rack for 10–15 min.
2. Thermo-regulated growth room/cabinet:
   Cell culture and calli formation need to be incubated in a dark thermo-regulated chamber or cabinet at +24–25 °C with ventilation to homogenize temperature (*see* **Note 1**). The oxygenation of the cell suspensions is maintained by constant agitation using orbital shakers such as Orbit 1900 (Labnet, USA) (*see* **Note 2**). Liquid *Agrobacterium tumefaciens* cultures for stable transformation of plant cell cultures are incubated at +28 °C with agitation at 180 rpm.
3. Washing and harvesting cell suspension cultures:
   The washing of cell suspensions to calibrate cell density, to remove cell debris or to wash compounds/excess *Agrobacterium* is performed by using a swing-out rotor with bucket adaptors for 15 and 50 mL conical tubes such as an Eppendorf centrifuge 5804R with A-4-44 rotor. The harvesting of cell samples is performed either by centrifugation (*see* **Note 3**) or by vacuum filtration (such as the Sterifil system from Millipore, USA) through two layers of 70 μm nylon mesh (*see* **Note 4**).
4. Culture flasks and containers:
   Cell suspensions are incubated in sterile containers which depend on the volume of culture needed being either manufacturer's individually sterilized multi-well plates or autoclaved

Erlenmeyer conical glass flasks. Usually 2–4 mL of cell suspensions are transferred per well in six-well plates, 10–20 mL in 50 mL Erlenmeyer flasks, 30–100 mL in 250 mL Erlenmeyer flasks (Fig. 1) and 200–500 mL in 3 L Erlenmeyer flasks depending on the amount of material needed (*see* **Note 5**).

### 2.2 Cell Culture Media, Hormones and Solutions

1. Surface sterilization solution:
   Surface sterilization solutions for seeds or explants include 70% ethanol, autoclaved ultrapure water and 5% commercial bleach in water (Clorox, Oakland, USA) while surface sterilization of external non-sterile equipment and user's hands/gloves is performed with 70% ethanol in spray bottles (*see* **Note 6**).
2. Murashige & Skoog (MS) culture medium:
   Heterotrophic *Arabidopsis thaliana* calli and cell suspensions are grown on 1× MS medium with 3% sucrose at pH 5.7. Culture medium is made in 900 mL of ultrapure water magnetically stirred in a 1 L graduated beaker to which 30 g of sucrose (27480, VWR chemicals) and 4.4 mg MS salt including vitamins (M0222, Duchefa) are added. The solution is adjusted to pH 5.7 with 2–3 M KOH solution and filled up to 1 L with ultrapure water. Jellified media are made by supplementing medium solution with either 0.8% (w/v) Agar-agar (1.01614, Merck) or 0.4% (w/v) phytagel (F8164, Sigma-Aldrich). Culture media are then autoclaved: jellified media are poured into sterile petri plates, sealed with Parafilm and stored at 4 °C while liquid media are stored at room temperature (*see* **Note 7**).
3. TE induction medium:
   *Arabidopsis thaliana* TE differentiation medium is composed of 1× MS medium with 3% sucrose and 1 mM 2-(N-morpholino) ethanesulfonic acid (MES) (M5287, Sigma-Aldrich) at pH 6.0. Medium is prepared as above in **item 2**, and 195 mg of MES are added prior to adjustment to pH 6.0 with 2–3 M KOH. Culture media are then autoclaved and stored at room temperature (*see* **Note 7**).
4. Luria-Bertani (LB) medium for *Agrobacterium tumefaciens*:
   5× LB bacterial growth medium is made by dissolving 50 g tryptone, 25 g yeast extract and 50 g NaCl in 900 mL of ultrapure water, adjusted to pH 7.0 with 2 M NaOH and completed to 1 L with ultrapure water. 1× LB liquid solution are made by combining 1 volume of 5× concentrated LB solution with 4 volume of sterile autoclaved ultrapure water. Jellified media are made by supplementing 1× LB solution with 1.5% (w/v) agar-agar (1.01614, Merck). Culture media are then autoclaved: jellified media are poured into sterile petri plates and stored at +4 °C while liquid media are stored at room temperature (*see* **Note 7**).

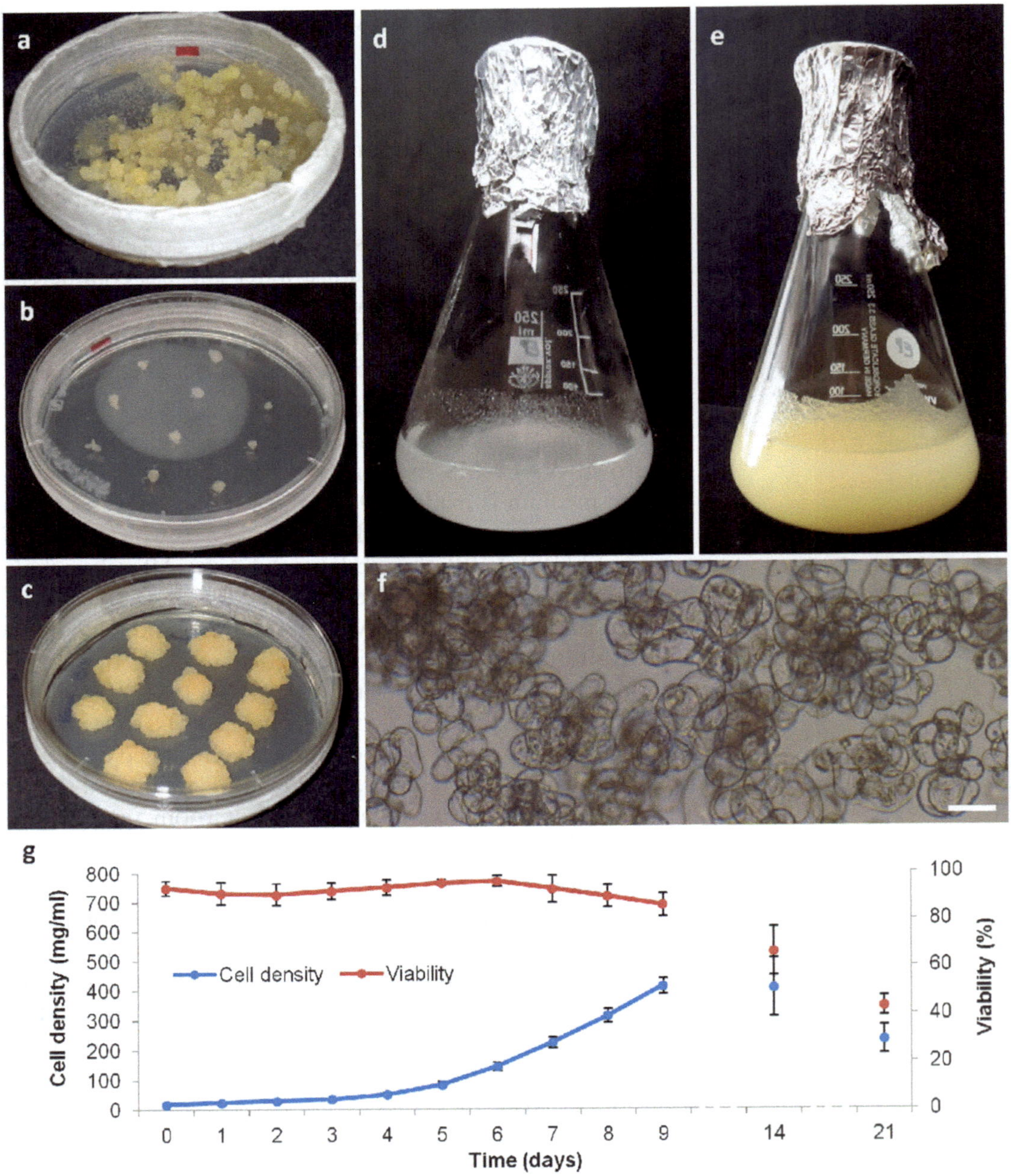


**Fig. 1** Standard procedures for calli transfer to solid and liquid media during establishment, screening for habituation, titration and screening of stable transgenic transformation. (**a**) Callogenesis on solid medium after cocultivation of habituated cells with *Agrobacterium tumefaciens* in liquid selective medium. (**b**) Calli placed onto fresh petri dish for growth using sterile tweezers for transfer. (**c**) Growth of calli after 6 weeks of incubation prior to liquid transfer. (**d**) Freshly subcultured habituated cell suspension at 60 mg/mL in sterile 250 mL Erlenmeyer flask. (**e**) 7-day-old habituated cell suspension in sterile 250 mL Erlenmeyer flask. (**f**) Appearance of grouped and isolated cells in the habituated cell suspensions at 7-days of culture. (**g**) Progression of cell density and cell viability—expressed respectively in mg of fresh weight/mL of cell suspension and in percentage of all cells present—during the growth time-course. Error bars indicate standard deviation (n=6 replicates)

5. Growth and inductive hormone solutions:
Hormonal solutions include 1000× concentrated solution of 6-benzylaminopurine (BAP) (B3408, Sigma-Aldrich) at 1 mg/mL, 166× concentrated solution of 1-naphthaleneacetic acid (NAA) (N0640, Sigma-Aldrich) at 10 mg/mL, 1000× concentrated solution of 2,4 dichlorophenoxyacetic acid (2,4D) (D7299, Sigma-Aldrich) at 1 mg/mL as well as 1200× concentrated solution of epibrassinolide (E1641, Sigma-Aldrich) at 5 mM. The solution preparation depends on the hormone: (1) 10 mg of BAP or 100 mg of NAA are placed in sterile 15 mL conical tubes with 0.5 mL of 2–3 M KOH, vortexed until completely dissolved and filled up to 10 mL with ultrapure water; (2) 10 mg of 2,4D are transferred to a 15 mL sterile conical tube to which 1 mL of 99% ethanol is added, vortexed until complete dissolution and adjusted to 10 mL with sterile ultrapure water; and, (3) 10 mg of epibrassinolide are solubilized in 1 mL dimethylsulfoxide (DMSO), vortexed and transferred to a 15 mL sterile conical tube. This step is repeated with DMSO until a final volume of 4.2 mL is reached. Sterile hormone solutions are filtered through 0.2 μm syringe filters and distributed as 0.8–1 mL aliquots in 1.5 mL autoclaved microtubes. All solutions are stored at −20 °C.
6. Antibiotic and herbicide solutions:
Compounds for transgenic culture selection and bacteria/plant cell growth arrest include 1000× concentrated solution of kanamycin monosulfate (K0126, Duchefa) at 50 mg/mL in ultrapure water (w/v), 2000× solution of hygromycin B at 20 mg/mL (A2175, Applichem) in ultrapure water (w/v), 500× of spectinomycin dihydrochloride solution at 50 mg/mL (S0188, Duchefa) in ultrapure water (w/v), 2000× solution of glufosinate herbicide (Basta, Bayer CropScience) at 20 mg/mL in ultrapure water (w/v) and 500× solution of cefotaxime sodium at 200 mg/mL (C0111, Duchefa) in ultrapure water (w/v). Solutions are prepared by weighing 50 mg of kanamycin, 50 mg of spectinomycin or 2 g of cefotaxime in sterile 15 mL conical tubes, adjusted and vortexed until completely dissolved in 10 mL of ultrapure water. Hygromycin (1/2 dilution of Hygromycin B stock) and glufosinate (1/10 dilution of Basta stock) are diluted from commercially available liquid concentrated stock solutions with ultrapure water in 15 mL sterile conical tubes. All solutions are sterilized through 0.2 μm syringe filters and stored at −20 °C except for glufosinate which is stored at +4 °C.
7. Pharmacological compound solutions:
Pharmacological compounds are solubilized in ultrapure water or dimethylsulfoxide (DMSO) and are sterilized through 0.2 μm syringe filters and stored at −20 °C. Generally,

1000× concentrated stock solutions are made when compound solubility allows it.

8. Staining solutions to monitor the progression of TE differentiation:
   Viability staining solutions enable the monitoring of living cells and include (1) 0.5 % (w/v) solution of fluorescein diacetate (FDA) (F7378, Sigma-Aldrich) solubilized in 100 % acetone [25] or (2) brigthfield colorigenic staining with 1 % (w/v) solution of Evans Blue (EB) (E2129, Sigma-Aldrich) solubilized in ultrapure water. Cell wall cellulose staining solution is composed of 0.01 % (w/v) calcofluor/fluorescent brightener 28 (F3543, Sigma-Aldrich) solubilized first in 1 mL of 2–3 M KOH and then completed with ultrapure water. All staining solutions are vortexed and filtered through 0.45 μm syringe filters. All solutions are protected from light: FDA solution is stored at −20 °C, while EB and calcofluor solutions are kept at room temperature.

## 3 Methods

All steps of the cell culture preparations and manipulations are to be performed under a sterile laminar flow hood (*see* **Note 6**).

### *3.1 Callogenesis from Explants*

1. Explant surface sterilization
   Explants—seeds or plant sectioned organ—are surface sterilized by placing them in sterile plastic containers (microtubes or conical tubes) and by sequentially washing the plant samples with (1) sterile water, two times for 5 min, (2) 70 % ethanol, once for 2 min, (3) sterile water, two times for 5 min, (4) 5 % Clorox solution, once for 5 min and (5) sterile water, two times for 5 min. Seeds are placed in square petri dishes filled with 0.5× MS medium with 0.8 % agar-agar, no sugars at pH 5.7 using pipette and sterile tips. Plates are then kept at +4 °C for at least 2 days for seed stratification and incubated vertically under long day conditions with 16/8-h day/night photoperiod, +22 °C/+18 °C temperature, and 70 % relative humidity.
2. Callus induction
   Explant fragments (2–3 cm long)—either directly after surface sterilization or sectioned with sterile tweezers from in vitro grown seedlings—are transferred to the callus induction medium composed of 1× MS medium with 3 % sucrose and 0.8 % agar-agar (or 0.4 % Phytagel) at pH 5.7 and supplemented with 0.5 mg/L of 2,4D and 0.05 mg/L of BAP. Samples are then incubated for 3–6 weeks in the dark at +24–25 °C (*see* **Notes 1** and **10**). Calli generally form at the extremities of explants and damaged areas caused during the handling with tweezers (Fig. 1a and b).

3. Calli transfer and growth
   Calli (diameter > 2 mm) are separated from the explants and transferred using sterile tweezers to callus growth medium composed of 1× MS medium with 3% sucrose and 0.8% agar-agar (or 0.4% Phytagel) at pH 5.7 and supplemented with 1 mg/L of 2,4D (Fig. 1b). Samples are then incubated for 3–6 weeks in the dark at +24–25 °C (*see* **Note 1**) (Fig. 1c).
4. Liquid resuspension of calli
   Calli (diameter > 1 cm) are severed in two and half is transferred using sterile tweezers in six-well sterile microplates containing 2–3 mL of 1× MS medium with 3% sucrose at pH 5.7 supplemented with 1 mg/L of 2,4D. Calli are fragmented using mechanical severing with sterile tweezers as well as carefully pipetting up and down the cell clumps. Samples are then incubated for 1–2 weeks with agitation (120 rpm for 1.9 cm orbits) in the dark at +24–25 °C and monitored every day (Fig. 1d and e). Samples exhibiting active growth are transferred to sterile Erlenmeyer flasks and the cell culture volume is gradually increased by 50% every 1–2 weeks (*see* **Note 11**).
5. Removing cell debris
   Growing cell and calli fragments with browning are washed by transferring the growing cells into a sterile conical tubes and gentle centrifugation (200 *g* for 2 min at room temperature) using a swing out rotor centrifuge (*see* **Note 3**). Extracellular medium is removed and replaced with fresh 1× MS medium with 3% sucrose at pH 5.7. Cell pellets are gently resuspended by inverting the tube several times (until no pellet remains at the conical end of the tube) and subjected to two additional rounds of centrifugation and medium removal/resupply. The last medium to be supplied is 1× MS medium with 3% sucrose at pH 5.7 supplemented with 1 mg/L of 2,4D.
6. Cell subculturing
   10–50% 7-day-old cell suspensions are weekly subcultured into fresh 1× MS medium with 3% sucrose at pH 5.7, supplemented with 1 mg/L of 2,4D and incubated with agitation (120 rpm for 1.9 cm orbits) in the dark at +24–25 °C (*see* **Notes 1** and **2**) (Fig. 1d and e).

### 3.2 Cell Suspension and Screening for Habituation

Habituated growth of plant cell suspension cultures has been observed at the onset of the technical development of in vitro plant cell cultures as Gautheret noticed in 1942 that certain established carrot cell lines gradually lose their requirement for auxin, cytokinins and vitamins [26]. Habituation in plant cell in vitro culture was thus noticed for many plant species including *Beta vulgaris* [27], *Arabidopsis thaliana* [28], *Catharanthus roseus* [29], *Helianthus annuus* [30], *Vinca rosea* [30], and *Daucus carota* [31]. Growth habituation has been attributed to epigenetic changes leading to the activation of transposable elements [32]; however, habituated cell lines still keep the capacity to respond to external factors such as hormones—observed for *Beta vulgaris* [27] and *Arabidopsis thaliana* [18]—and/or composition of the medium such as the type of sugar—as observed in *Arabidopsis thaliana* between sucrose, glucose, and fructose [33].

1. Screening for habituated growth of cell suspension culture
   2–3 L of 7-day-old cell suspension culture (Fig. 1f) are spread onto petri dishes containing 1× MS medium with 3% sucrose and 0.8% agar-agar (or 0.4% Phytagel) at pH 5.7 and incubated in the dark at +24–25 °C for 10–18 weeks. Habituated calli (diameter > 2 mm) are then transferred onto fresh solid medium (1× MS medium with 3% sucrose and 0.8% agar-agar (or 0.4% Phytagel) at pH 5.7) using sterile tweezers and incubated 3–6 weeks in the dark at +24–25 °C (*see* **Note 1**) (Fig. 1a).
2. Transfer of calli to liquid medium
   Calli (diameter > 1 cm) are severed in two and half is transferred using sterile tweezers in a six-well sterile microplate containing 2–3 mL of 1× MS medium with 3% sucrose at pH 5.7. Calli are fragmented using mechanical severing with sterile tweezers and carefully pipetting up and down. Samples are then incubated for 1–2 weeks with agitation (120 rpm for 1.9 cm orbits) in the dark at +24–25 °C and monitored every day. Samples exhibiting active growth are transferred to sterile Erlenmeyer flasks and the cell culture volume is gradually increased by 50% every 1–2 weeks (Fig. 1d–g).
3. Weekly maintenance of habituated cell line
   10–20% 7-day-old cell suspensions are weekly subcultured (*see* **Note 11**) into fresh 1× MS medium with 3% sucrose (or another sugar source) at pH 5.7 and incubated with 120 rpm agitation in the dark at +24–25 °C (*see* **Notes 1** and **2**) (Fig. 1d–g).
4. Titration of habituated cells
   Cell growth habituation can be either gained or lost during successive cell subculturing through time thus leading to chimeric cell cultures. Therefore, habituated cell suspensions are routinely titrated for habituation by plating 2–3 mL of 7-day-old cell at 1/2, 1/5, and 1/10 dilution with fresh 1× MS medium with 3% sucrose at pH 5.7 onto petri dishes containing 1× MS medium with 3% sucrose and 0.8% agar-agar (or 0.4% Phytagel) at pH 5.7 and incubated 3–6 weeks in the dark at +24–25 °C. Resulting calli can be transferred back to solid medium as in **step 1** of Subheading 3.2 or to liquid medium as in **step 2** of Subheading 3.2 (Fig. 1a–c).

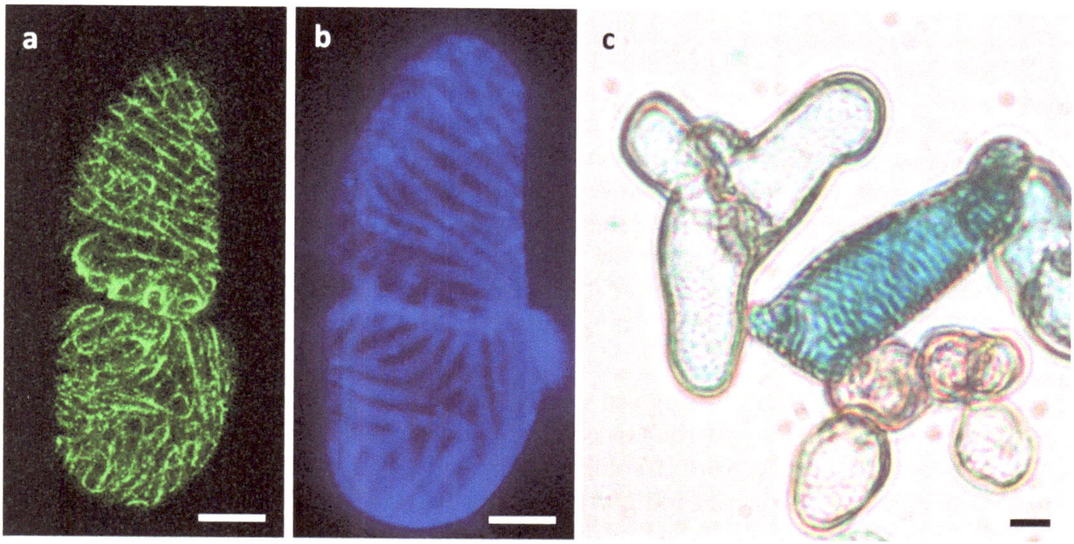


**Fig. 2** Examples of the multiple types of stable transgenic transformation of *Arabidopsis thaliana* habituated cell suspension cultures. (**a**, **b**) Image of induced TE from habituated cell suspension cultures of *Arabidopsis thaliana* stably transformed with constitutively expressed MAP70-5 fused with GFP (**a**) and corresponding calcofluor-stained cell wall (**b**), bar = 10 μm. (**c**) Image of induced TE and parenchymatic cells from habituated cell suspension cultures of *Arabidopsis thaliana* stably transformed with promoter:GUS reporter construct, bar = 10 μm

### *3.3 Stable Genetic Transformation of Habituated Cell Suspensions*

Stable transgenic transformation of habituated cell suspension cultures are performed using horizontal gene transfer by coculture with transgenic *Agrobacterium tumefaciens* exhibiting both the helper plasmid and the modified gene construct to be stably randomly inserted (Fig. 2). All steps of plant cell suspension stable genetic transformation are performed under sterile conditions (*see* Subheading 2.1). This method for genetic transformation is quite stable in time as transgenic cell suspension cultures are still capable of TE differentiation and GFP fusion expression 7 years after establishment [18–22].

1. Coculture of *Arabidopsis* habituated cells with *Agrobacterium tumefaciens*

   5 mL of 3–5-day-habituated plant cell suspensions are transferred to a 50 mL Erlenmeyer flask, supplemented with 5 mL of fresh 1× MS medium with 3 % sucrose at pH 5.7 and 200 μL of 2 days grown transgenic *Agrobacterium tumefaciens* (cultivated in 1× LB medium with selection antibiotics like spectinomycin until O.D. 600 nm ~1). The cocultures are incubated for 2 days in the dark with agitation at +24–25 °C. Both LB4404 and GV3101 strains have been used successfully [18, 22].
2. Removal of excess *Agrobacterium tumefaciens* and damaged plant cells

The coculture is transferred to a 15 mL Falcon conical tube and gently pelleted at 200 *g* for 2 min at room temperature using a swing out rotor centrifuge (*see* **Note 3**). The extracellular medium is removed delicately without disturbing the cell pellet (generally the pellet is between 1–2 mL packed cell volume) and supplemented with 10 mL of fresh 1× MS medium with 3% sucrose at pH 5.7. Cells are resuspended in the fresh medium by inverting the tube several times (until no pellet remains at the conical end of the tube) and subjected to three additional rounds of centrifugation and medium removal/resupply.

3. Plating cell suspension on selective medium
At the last centrifugation step, the cell pellet is supplemented with a volume of fresh 1× MS medium with 3% sucrose at pH 5.7 corresponding to 1.5–2 times the current packed cell volume in the conical tube and resuspended by gently pipetting up and down. 2–3 mL of resuspended cells are then placed onto the surface of petri plates filled with 1× MS medium with 3% sucrose and 0.8% agar-agar (or 0.4% Phytagel) at pH 5.7 supplemented with 200–250 μg/mL of cefotaxime and selective antibiotics/herbicides depending on the construct used—either 50 μg/mL of kanamycin, 5 μg/mL of hygromycin or 20 μg/mL of Basta herbicide. Plates are then incubated for 5–12 weeks in the dark at +24–25 °C (Fig. 1a).

4. Transgenic calli selection
Clearly separated calli (diameter > 2–3 mm) from the selection plates are then transferred using sterile tweezers onto the surface of new petri plates filled with 1× MS medium with 3% sucrose and 0.8% agar-agar (or 0.4% Phytagel) at pH 5.7 supplemented with 200–250 μg/mL of cefotaxime and selective antibiotic/herbicide markers. Plates are then incubated for 2–5 weeks in the dark at +24–25 °C (Fig. 1b and c).

5. Liquid transfer of calli and screening of transgenic lines
Freshly developed calli are broken into pieces (0.5–0.7 cm diameter) and placed in 2–4 mL of fresh liquid 1× MS medium with 3% sucrose at pH 5.7 with selective marker and incubated in sterile six-wells plate or 25 mL Erlenmeyer flasks for 1 week in the dark with agitation at +24–25 °C (*see* **Note 2**) (Fig. 1d and e). The screening of the generated lines is then performed according to the genetic construct stably inserted such as the overexpression of GFP fusion proteins (Fig. 2a and b) or promoter driving β-glucuronidase (GUS) reporter gene (Fig. 2c).

### 3.4 Tracheary Element Hormone-Dependent Differentiation

The hormonal induction of parenchymatic habituated cell suspension culture into sclerenchymatic TEs is highly dependent on (1) the cell density, (2) the inductive medium and (3) the hormone concentration/proportion as previously illustrated in multiple TE inductive systems from parenchymatic cells of *Zinnia elegans* [13], *Lactuca sativa* [8], *Helianthus tuberosus* [34] as well as in *Arabidopsis thaliana* [14, 15, 18].

1. Cell density adjustment
   10 mL of habituated cell suspensions at the plateau growth phase (9–11 days after subculturing) (Fig. 3a) are transferred to a sterile 15 mL Falcon conical tube (weighed empty beforehand) and gently pelleted at 200 ×*g* for 2 min at room temperature using a swing out rotor centrifuge. Under sterile conditions, the extracellular medium is then carefully removed using a controllable suction pipette (i.e., using Eppendorf pipetboy) until the pellet is not covered by any medium and weighted. The cell pellet is then resuspended in 5 mL fresh 1× MS medium with 3% sucrose and 1 mM MES at pH 6.0 and inverted until no pellet remains at the conical end of the tube. Newly resuspended cells are then transferred to a 250 mL Erlenmeyer flask and supplemented with fresh 1× MS medium with 3% sucrose and 1 mM MES at pH 6.0 until final cell density reaches 30 mg of cells/mL of medium (final volume depends on the pellet weight—*see* **Note 12**).
2. Hormonal induction
   Newly readjusted cell suspensions to a density of 30 mg of cells /mL are then induced to differentiate into TEs by adding 6 μg/mL of NAA (0.6 μL of the 166× stock /mL of culture medium), 1 μg/mL of BAP (1 μL of the 100× stock/mL of culture medium) and 4 μM of epibrassionolide (0.8 μL of the 1200× stock/mL of culture medium) and further incubated in the dark, under agitation at +24–25 °C for up to 3 weeks if needed. Any changes to the cell age (Fig. 3a), cell density (Fig. 3i), pH of the medium (Fig. 3k) and the hormone concentrations (Fig. 3j and l) will lead to a reduction of TE differentiation efficiency [37]. At least two samples are required to monitor TE differentiation: (i) induced cells supplemented with the hormones and (ii) control cells without any hormones added, both with an initial cell density of 30 mg/mL in separate Erlenmeyer flasks.
3. Sampling of cells during the TE differentiation time course
   During TE differentiation, cell suspensions are harvested by transferring the cell suspensions to a conical tube either (1) by centrifugation at 200 ×*g* for 5 min at room temperature using a swing-out rotor and removing the excess culture medium by pipetting (*see* **Note 13**) or (2) by vacuum-filtration through a double 70 μm nylon mesh and transferring the cells with a clean spatula to an aluminum foil folded sheet. The resulting cell pellets or filtrates are immediately flash frozen in liquid nitrogen and stored at −80 °C.

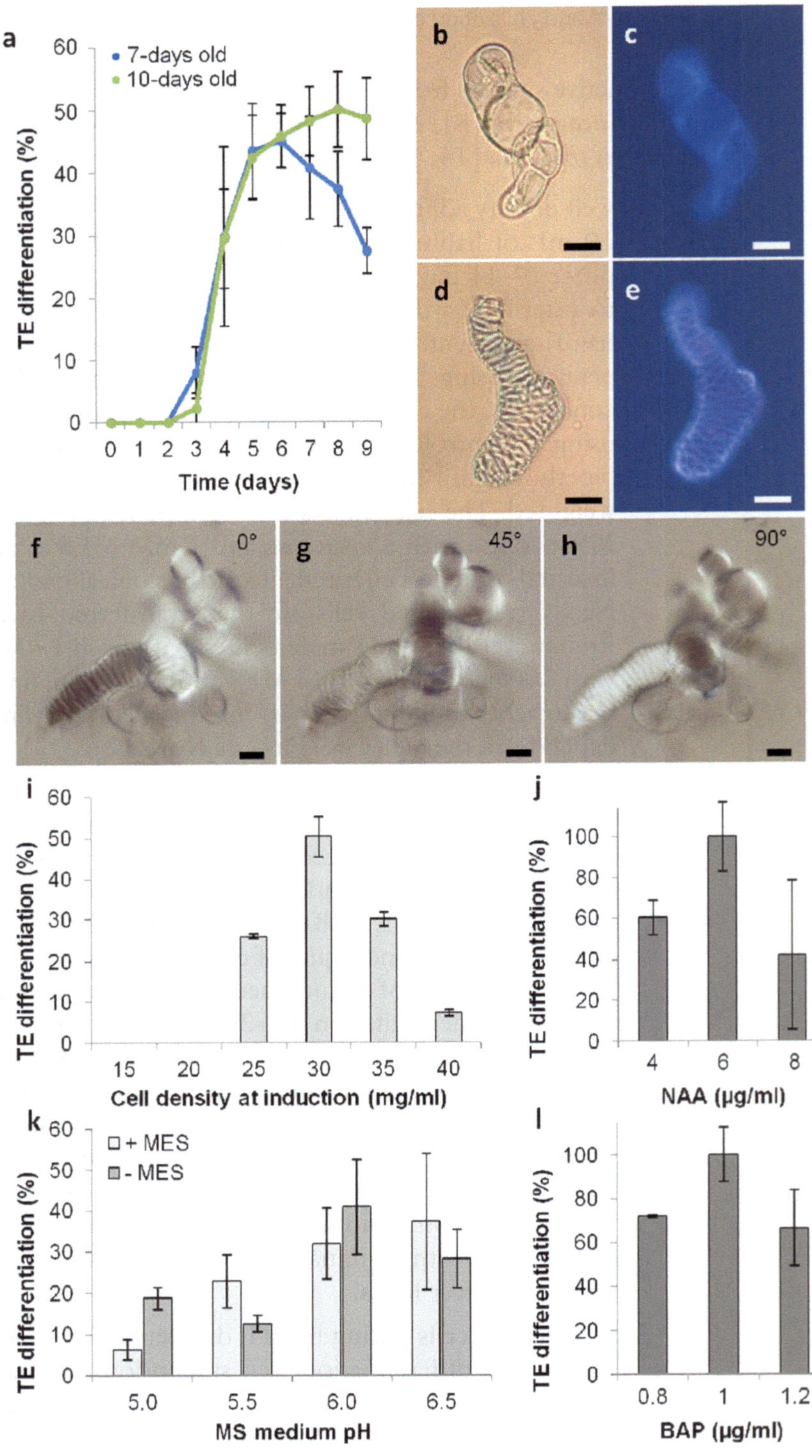


**Fig. 3** Factors influencing the efficiency of TE differentiation in habituated cell suspension cultures of *Arabidopsis thaliana*. (**a**) Effect of the cell culture age (7 days or 10 days after subculture into fresh medium) on the progression of TE differentiation—expressed in percentage of TEs among all cells present—during the weekly growth time-course. (**b**) Initial undifferentiated parenchymatic cell, bar = 10 μm. (**c**) Calcofluor-stained undifferentiated parenchymatic cell, bar = 10 μm. (**d**) Individual TE produced in habituated cell suspensions, bar = 10 μm. (**e**) Calcofluor-stained TE produced in habituated cell suspensions, bar = 10 μm. (**f–h**) Birefringence of TE secondary cell wall using polarized microscopy with stage orientated at 0° (**f**), 45° (**g**) and 90° (**h**), bars = 10 μm. (**i**) Effect of cell density—expressed in mg of fresh weight per mL—on TE differentiation efficiency (expressed in

### 3.5 Monitoring Cell Culture Growth and TE Differentiation

Single cell systems enable both the direct microscopic analysis of the individual cell state in the entire suspensions as well as the characterization of the global population state during the time-course of TE differentiation. The monitoring of cell density and viability needs to be performed for the different steps to establish habituated cell suspensions cultures including **step 6** of Subheading 3.1, **step 3** of Subheading 3.2, **step 5** of Subheading 3.3, **step 3** of Subheading 3.4.

1. Cell growth/density
   The measurements of cell density in suspension cultures during the differentiation time-course can be estimated by: (1) the *packed cell volume* (PCV) which corresponds to the volume of cells pelleted by 2 min centrifugation at $200 \times g$ for 10 mL of suspension using a graduated conical tube [33]; (2) by the fresh weight per volume of cell suspension which similarly estimates the weight of cells obtained after vacuum filtering 10 mL of cell suspensions (Fig. 1a) [33]; and (3) the dry weight per volume of cell suspension which estimates the weight of cells after vacuum filtering 10 mL of cell suspensions, transferred to an aluminum sheet and dried in an oven overnight at +70 °C [33].
2. Cell viability
   The estimation of cell viability at the single cell level during the differentiation time-course is estimated either by (1) fluorescent staining using 2 μL of 0.5 % FDA solution for 100 μL of cell suspension [25] or (2) brigthfield colorigenic staining using 1:1 (v/v) incubation with 1 % EB solution. Cells are incubated 2–5 min at room temperature and observations are performed using light microscopy by placing 50 μL of the stained samples between microscopy glass slides and coverslips: EB staining colors dead cells in blue in brightfield transmission (*see* **Note 8**); FDA staining colors living cells with fluorescein (green fluorescence) which is observed using blue light excitation (excitation filter 450–490 nm; emission filter 515 nm) (*see* **Note 9**) [25]. Around 500–600 individual cells are counted to obtain representative results. The proportion of living cells can be expressed either relatively to all cells in the population or to the proportion of visible TEs or other cell types.
3. TE differentiation
   The measurement of TE differentiation efficiency at the single cell level during the differentiation time-course is estimated by

**Fig. 3** (continued) percentage of TEs among all cells present). (**j**) Effect of changes in NAA concentration on TE differentiation efficiency (expressed as percentage to control set at 6 μg/mL) for 1 μg/mL BAP and 4 μM epibrassinolide. (**k**) Effect of medium pH and the presence of MES on TE differentiation efficiency (expressed in percentage of TEs among all cells present). (**l**) Effect of changes in BAP concentration on TE differentiation efficiency (expressed as percentage to control set at 1 μg/mL) for 6 μg/mL NAA and 4 μM epibrassinolide

(1) brigthfield microscopy due to the birefringence of TE secondary cell walls (Fig. 3f–h) [22], (2) fluorescent staining of cellulose using 10 μL of 0.01 % calcofluor solution for 100 μL of cell suspensions (Fig. 3b–e) [18, 25] or (3) autofluorescence of lignin [38]. Cells are incubated 2–5 min at room temperature and observations are performed using light microscopy by placing 50 μL of the stained samples between microscopy glass slides and coverslips. Calcofluor-stained cells, which can be combined with FDA, are observed with UV excitation light (excitation filter 270–380 nm; emission filter 410–580 nm). TE lignin autofluorescence is visualized without any staining with UV excitation light (excitation filter 270–380 nm; emission filter 410–580 nm) [38]. Around 500–600 individual cells are counted to obtain representative results. The TE differentiation efficiency is expressed relatively to all cells in the population (Fig. 3a).

### 3.6 Live Cell Imaging of TE Differentiation

Due to the single cell state of habituated *Arabidopsis thaliana* cell suspension cultures, the complete differentiation process of TEs can readily be observed using real time live-cell imaging which has enabled to determine the duration of TE xylan/cellulose secondary cell wall deposition [18, 23], the duration of nuclear degradation [5, 18, 23] and vacuole implosion [5] during TE programmed cell death as well as the timing of lignin deposition during TE differentiation [18, 22].

1. Sample mounting in microscopy bio-chamber
   200 μL of cell suspension culture are transferred to the bio-chamber (one for the control and one for the treatment), onto the poly-lysine coverslip delimiting the bottom part of the 3 mm thick acrylic plastic bio-chamber. Design and assembly of the bio-chamber have been described as a detailed protocol in [35]. The cell suspension is spread all over the coverslip surface and excess culture medium and non-adherent cells are removed by pipetting. This step is repeated twice to fully saturate the poly-lysine surface with cells.
2. Immobilization of cells in microscopy bio-chamber
   500 μL of 0.5 % low melting point agarose in 1× MS medium are slowly added from one corner of the well until a meniscus is formed at the top of the well. Upper surface is then flattened for observation using another coverslip to temporally seal the bio-chamber. Medium is left for 10–15 min to set so that the low melting point agarose jellifies and cells are unable to freely move in the chamber.
3. Mounting of the bio-chamber on the upright microscope
   Initially the sealing coverslip is removed delicately without damaging the agarose surface and replaced with a microscope slide separated from the bio-chamber by two spacing strips of

Whatmann paper 3 mm (3–4 mm wide). The assembled biochamber is then mounted in the holding brackets of the upright microscope (i.e., Zeiss Aixoplan II with automated *xyz* stage equipped with Axiocam camera and controlled with Axiovision software) for real-time live cell imaging observations (Fig. 4). Real-time acquisition is then performed as described in the detailed protocol [35].

### 3.7 Pharmacological Modulation of TE Differentiation

The semi-synchronous differentiation of TEs in cell suspension cultures enables to treat or wash out any pharmacological compounds from the cells at a specific time-point during the differentiation process. Delayed treatments and/or washing-outs allow to specifically determine when and which of the sequential process is affected by the pharmacological compound as previously performed for the ethylene biosynthetic inhibitor aminovinylglycine (AVG) [25], the ethylene perception inhibitor silver thiosulfate (STS) [5, 36] and the microtubule destabilizing compounds oryzalin (Fig. 4a–d) [22, 24].

1. Delayed pharmacological treatments (can be performed on both six-well microplates and larger Erlenmeyer flasks): the pharmacological compound treatments are made at a defined time after the hormonal induction of TE differentiation. One example is the impact of the delayed oryzalin treatment on TE secondary cell wall organization [22].
2. Removal of the pharmacological inhibition by washing (can be performed on both six-well microplate and larger Erlenmeyer flasks): cells are transferred to a 15 or 50 mL Falcon conical tube and gently pelleted at 200 × *g* for 2 min at room temperature using a swing out rotor centrifuge. The total volume is recorded and the extracellular medium is then removed delicately without disturbing the cell pellet and replaced with the same volume of fresh TE induction medium (1× MS medium with 3% sucrose and 1 mM MES at pH 6.0). Cells are resuspended in the fresh medium by inverting the tube and subjected to two additional rounds of centrifugation and medium removal. Cells are then incubated by placing them either into fresh medium or the extracellular medium from cells at the same developmental stages which were not treated with the pharmacological compounds.

## 4 Notes

1. Note that white light contamination will affect the growth and differentiation of habituated cell suspensions, green safety light is therefore recommended for handling cells in the growth room. Additionally, black painted wall and a safety lock delimited by

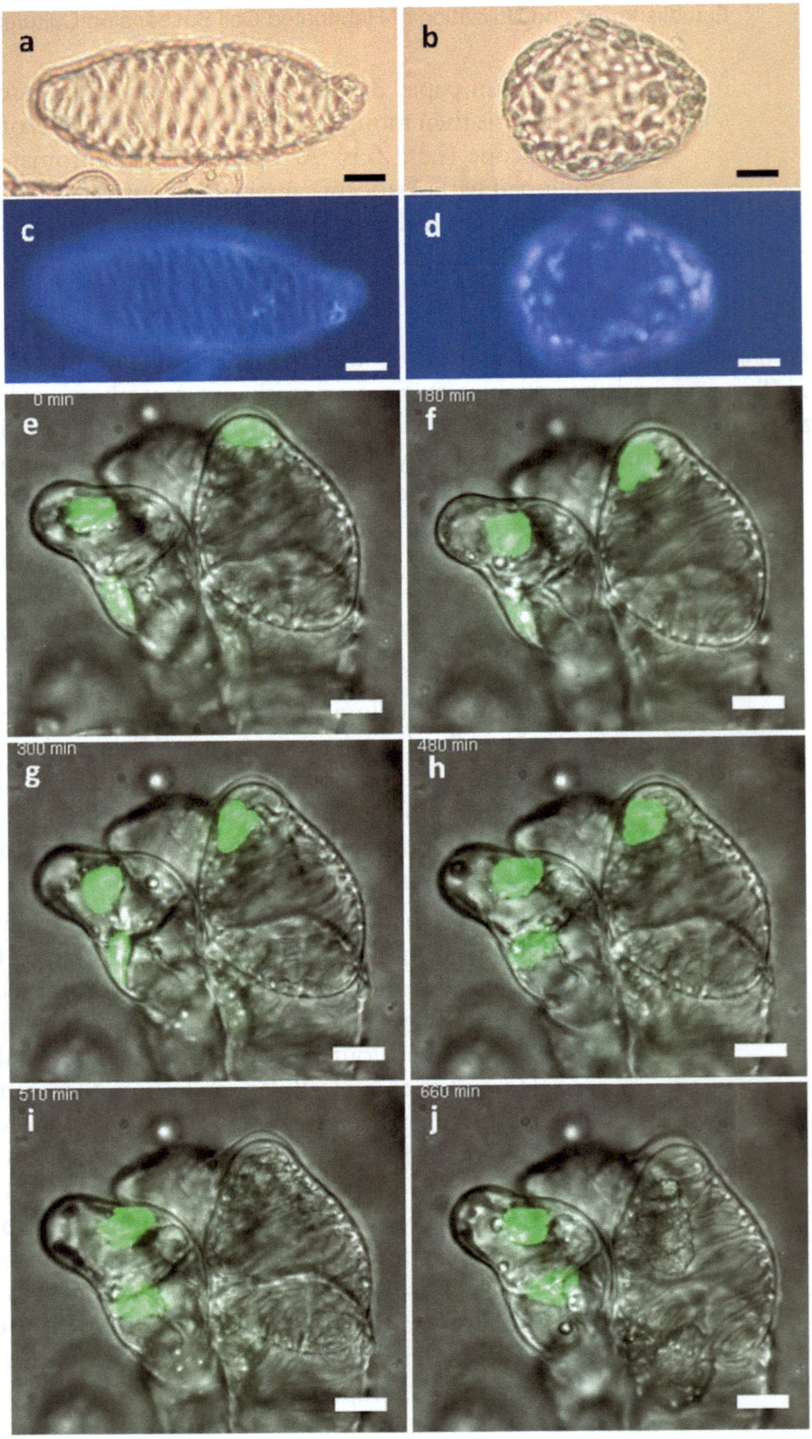


**Fig. 4** Examples of pharmacological modifications and live-cell imaging during TE formation in habituated cell suspension cultures of *Arabidopsis thaliana*. (**a**) Image of control TE treated with DMSO observed in brightfield, bar = 10 μm. (**b**) Image of TE treated with 10 μM of orzyalin observed in brightfield, bar = 10 μm. (**c**) Image of control TE treated with DMSO stained with calcofluor, bar = 10 μm. (**d**) Image of TE treated with 10 μM of orzyalin stained with calcofluor, bar = 10 μm. (**e–j**) Chronological live-cell imaging montage of stable transgenic habituated cells constitutively expressing histone2A fused with GFP incubated in a bio-chamber and induced to differentiate into TEs during secondary cell wall formation and programmed cell death (merge images of brightfield and nuclear GFP fluorescence), bars = 10 μm

black curtains to enter the growth room represent additional features limiting white light contamination in the cell culture growth room.

2. Note that orbital agitation speed (in rpm) depends on the rotation diameter of the orbital shaker: 120 rpm are used for effective non-damaging agitation using an Orbit 1900 (Labnet, USA) with a 1.9 cm rotation diameter.
3. Note that centrifugation of cell suspensions needs to be made at room temperature or close to the growth incubation temperature of +24–25 °C, heat-shocks will induce changes in the growth and sensitivity of the habituated cell suspensions.
4. Note that vacuum filtering is the most efficient way to remove excess water from the collected cell suspension samples.
5. Note that Erlenmeyer conical flasks used for cell culture need to be extremely clean as previous dried residues of cell culture will induce the browning of cell suspensions. We routinely do two rounds of washing—first hand wash with swab, non-ionic detergent and hot water and then machine wash at +60 °C with cleaning and degreasing agent.
6. Note that the external surface of containers and the user's arms are surface sterilized by spraying with a solution of 70% ethanol or using alcogel when appropriate.
7. Note that autoclaving will cause a slight drop of the pH in the MS medium due to thermal cracking of the sucrose into small organic acid and simpler sugars. Production thermal cracked derivatives from sucrose can be avoided by mixing 1 volume of autoclaved 2× MS solution at pH 5.7 without any sugar with 1 volume of 6% sucrose in water (w/v) at pH 5.7 sterilized through a 0.2 μm filter [33].
8. Note that EB exhibits affinity for cellulose and will stain TE secondary cell walls in blue.
9. Note that long incubation with FDA will lead to an extracellular release of fluorescein which will prevent efficient numeration of cell viability.
10. Note that the sealing of petri plates for calli production and growth is best made with micro-pore tape rather than Parafilm which becomes brittle and cracks.
11. Note that cell suspension cultures can be incubated for a maximum of 1 h on the bench without any agitation. Longer incubation will lead to hypoxia/anoxia which will result in the death of the cell suspensions and its browning.
12. The weight of the cell pellet is determined by subtracting the weight of the tube with dried pellet minus the weight of the empty tube. The volume of medium to be added to reach 30 mg/mL is calculated by dividing the pellet weight (in mg) by 30.

13. Note that access to flash frozen samples for large samples is hindered by the presence of the containing tube which needs to be removed by hammering. Mechanical extraction in the containing tubes using vortexing with tungsten carbide beads and freeze–thaw cycles only works for small cell pellets of 10–50 mg [36].

## Acknowledgment

This work was supported by Vetenskapsrådet VR (grant 2010-4620), the BioImprove FORMAS (to E.P.), the Berzelii Centre for Forest Biotechnology (to D.M. and E.P.), by the Kempe Foundation (Gunnar Öquist Fellowship to E.P.), by the Carl Tryggers Foundation and by the Stiftelsen för Strategisk Forskning (ValueTree) to E.P. We also thank the Swedish research council Vetenskapsrådet and the Swedish Governmental Agency for Innovation Systems Vinnova (UPSC Berzelii Centre) and Bio4 Energy (a strategic research environment appointed by the Swedish government).

### References

1. Fukuda H (1996) Xylogenesis: initiation, progression, and cell death. Annu Rev Plant Physiol Plant Mol Biol 47:299–325
2. Ye ZH (2002) Vascular tissue differentiation and pattern formation in plants. Annu Rev Plant Biol 53:183–202
3. Turner S, Gallois P, Brown D (2007) Tracheary element differentiation. Annu Rev Plant Biol 58:407–433
4. Ménard D, Pesquet E (2015) Cellular interactions during tracheary elements formation and function. Curr Opin Plant Biol 23:109–115
5. Ménard D, Escamez S, Tuominen H, Pesquet E (2015) Life beyond death: the formation of xylem sap conduits. In: Gunawardena A, McCabe P (eds) Plant programmed cell death, advances in plant biology. Springer, New York, pp 55–76
6. Vöchting H (1892) Über transplantation am Pflanzenkörper: untersuchungen zur physiologie und pathologie. Verlag Der H.Laupp'schen Buchnadlung, Tübingen
7. Hepler PK, Newcomb EH (1963) The fine structure of young tracheary xylem elements arising by redifferentiation of parenchyma in wounded coleus stem. J Exp Bot 14:496–503
8. Dalessandro G (1973) Hormonal control of xylogenesis in pith parenchyma explants of lactuca. Ann Bot 37:375–382
9. Jeffs RA, Northcote DH (1966) Experimental induction of vascular tissue in an undifferentiated plant callus. Biochem J 101:146–152
10. Kwon SI, Cho HJ, Jung JH, Yoshimoto K, Shirasu K, Park OK (2010) The Rab GTPase RabG3b functions in autophagy and contributes to tracheary element differentiation in Arabidopsis. Plant J 64:151–164
11. Yamagishi Y, Yoshimoto J, Uchiyama H, Nabeshima E, Nakaba S, Watanabe U, Funada R (2013) In vitro induction of secondary xylem-like tracheary elements in calli of hybrid poplar (Populus sieboldii × P. grandidentata). Planta 237:1179–1185
12. Yamagishi Y, Sato T, Uchiyama H, Yoshimoto J, Nakagawa R, Nakaba S, Kubo T, Funada R (2012) Tracheary elements that resemble secondary xylem in calli derived from the conifers, Torreya nucifera and Cryptomeria japonica. J Wood Sci 58:557–562
13. Fukuda H, Komamine A (1980) Establishment of an experimental system for the study of tracheary element differentiation from single cells isolated from the mesophyll of Zinnia elegans. Plant Physiol 65:57–60
14. Kubo M, Udagawa M, Nishikubo N, Horiguchi G, Yamaguchi M, Ito J, Mimura T, Fukuda H, Demura T (2005) Transcription switches for

protoxylem and metaxylem vessel formation. Genes Dev 19:1855–1860

15. Oda Y, Mimura T, Hasezawa S (2005) Regulation of secondary cell wall development by cortical microtubules during tracheary element differentiation in Arabidopsis cell suspensions. Plant Physiol 137:1027–1036
16. Oda Y, Iida Y, Kondo Y, Fukuda H (2010) Wood cell-wall structure requires local 2D-microtubule disassembly by a novel plasma membrane-anchored protein. Curr Biol 20: 1197–1202
17. Yamamura M, Wada S, Sakakibara N, Nakatsubo T, Suzuki S, Hattori T, Takeda M, Sakurai N, Suzuki H, Shibata D, Umezawa T (2011) Occurrence of guaiacyl/p-hydroxyphenyl lignin in Arabidopsis thaliana T87 cells. Plant Biotechnol 28:1–8
18. Pesquet E, Korolev AV, Calder G, Lloyd CW (2010) The microtubule-associated protein AtMAP70-5 regulates secondary wall patterning in Arabidopsis wood cells. Curr Biol 20:744–749
19. Blee KA, Wheatley ER, Bonham VA, Mitchell GP, Robertson D, Slabas AR, Burrell MM, Wojtaszek P, Bolwell GP (2001) Proteomic analysis reveals a novel set of cell wall proteins in a transformed tobacco cell culture that synthesises secondary walls as determined by biochemical and morphological parameters. Planta 212:404–415
20. Goué N, Mortimer JC, Nakano Y, Zhang ZN, Josserand M, Ohtani M, Dupree P, Kakegawa K, Demura T (2013) Secondary cell wall characterization in a BY-2 inductive system. Plant Cell Tiss Org Cult 115:223–232
21. Endo S, Pesquet E, Tashiro G, Kuriyama H, Goffner D, Fukuda H, Demura T (2008) Transient transformation and RNA silencing in Zinnia tracheary element differentiating cell cultures. Plant J 53:864–875
22. Derbyshire P, Ménard D, Green P, Saalbach G, Buschmann H, Lloyd CW, Pesquet E (2015) Proteomic analysis of microtubule interacting proteins over the course of xylem tracheary element formation in Arabidopsis. Plant Cell 27:2709–2726
23. Pesquet E, Korolev AV, Calder G, Lloyd CW (2011) Mechanisms for shaping, orienting, positioning and patterning plant secondary cell walls. Plant Signal Behav 6:843–849
24. Pesquet E, Lloyd CW (2011) Microtubules, MAPs and xylem formation. In: Liu B (ed) The plant cytoskeleton, advances in plant biology. Springer, New York, pp 277–306
25. Pesquet E, Tuominen H (2011) Ethylene stimulates tracheary element differentiation in Zinnia cell cultures. New Phytol 190:138–149
26. Meins F (1989) HABITUATION: heritable variation in the requirement of cultured plant cells for hormones. Annu Rev Genet 23: 395–408
27. Kevers C, Filali M, Petit-Paly G, Hagège D, Rideau M, Gaspar T (1996) Habituation of plant cells does not mean insensitivity to plant growth regulators. In Vitro Cell Dev Biol Plant 32:204–209
28. Pischke MS, Huttlin EL, Hegeman AD, Sussman MR (2006) A transcriptome-based characterization of habituation in plant tissue culture. Plant Physiol 140:1255–1278
29. Eilert U, Deluca V, Kurz WG, Constabel F (1987) Alkaloid formation by habituated and tumorous cell suspension cultures of Catharanthus roseus. Plant Cell Rep 6:271–274
30. Sogeke AK, Butcher DN (1976) The effect of inorganic nutrients on the hormonal requirements of normal, habituated, and crown-gall tissue. J Exp Bot 27:785–793
31. Smith DL, Krikorian AD (1990) Low external pH replaces 2,4-D in maintaining and multiplying 2,4-D-initiated embryogenic cells of carrot. Physiol Plant 80:329–336
32. Tanurdzic M, Vaughn MW, Jiang H, Lee TJ, Slotkin RK, Sosinski B, Thompson WF, Doerge RW, Martienssen RA (2008) Epigenomic consequences of immortalized plant cell suspension culture. PLoS Biol 6:2880–2895
33. Kunz S, Pesquet E, Kleczkowski L (2014) Functional dissection of sugar signals affecting gene expression in Arabidopsis thaliana. PLoS One 9:e100312
34. Phillips R, Dodds JH (1977) Rapid differentiation of tracheary elements in cultured explants of Jerusalem artichoke. Planta 135:207–212
35. Buschmann H, Sambade-Zumoffen A, Pesquet E, Lloyd CW (2010) Microtubule dynamics in plant cells. Methods Cell Biol 97:373–400
36. Pesquet E, Zhang B, Gorzsás A, Puhakainen T, Serk H, Escamez S, Barbier O, Gerber L, Courtois-Moreau C, Alatalo E, Paulin L, Kangasjärvi J, Sundberg B, Goffner D, Tuominen H (2013) Non-cell autonomous post-mortem lignification of tracheary elements in Zinnia elegans. Plant Cell 25: 1314–1328
37. Escamez S, André D, Zhang B, Bollhöner B, Pesquet E, Tuominen H (2016) METACASPASE9 modulates autophagy to confine cell death to the target cells during Arabidopsis vascular xylem differentiation. Biol Open 5:122–129
38. Serk H, Gorzsás A, Tuominen H, Pesquet E (2015) Cooperative lignification of xylem tracheary elements. Plant Signal Behav 10: e1003753

# Chapter 5

# Tissue Culture for Xylem Differentiation with Arabidopsis Leaves

Masato Saito, Alif Meem Nurani, Yuki Kondo, and Hiroo Fukuda

## Abstract

Culture systems combined with molecular biological approaches have identified various key factors regulating vascular differentiation, which makes a great contribution to development of vascular biology. Recently, we established a novel culture system for xylem cell differentiation in the model plant *Arabidopsis thaliana* (Arabidopsis), in which ectopic xylem cells can be induced throughout leaf disks within 3–4 days. Here we describe detailed procedures of the culture system from sample preparation to vascular cell induction culture.

**Key words** Culture system, Arabidopsis, Transdifferentiation, Reprogramming, Mesophyll, Procambium, Xylem

## 1 Introduction

Culture system is a powerful tool for analyzing cell differentiation in plants. In particular, vascular tissues are deeply located inside the plant body, making it difficult to trace their developmental process. Therefore, many research efforts have focused on the establishment of cell culture system for vascular cell differentiation [1–4]. While these cell cultures have contributed to the identification of key regulators, they have a difficulty in introducing mutants and markers for molecular genetic analyses.

Previously, we have reported that glycogen synthase kinase 3 proteins (GSK3s) play central roles in xylem cell differentiation, and then found that a specific inhibitor of GSK3s, bikinin, has an activity to induce ectopic xylem cell differentiation [5, 6]. Based on these findings we reported a novel tissue culture system that can induce ectopic xylem differentiation by the coculture with auxin, cytokinin, and bikinin using Arabidopsis leaves [7]. In this culture system, first mesophyll cells in leaves change their fates into procambial/cambial cells and then differentiate into xylem cells [7]. Since the Arabidopsis model system equips with the abundant

Miguel de Lucas and J. Peter Etchells (eds.), *Xylem: Methods and Protocols*, Methods in Molecular Biology, vol. 1544,
DOI 10.1007/978-1-4939-6722-3_5, © Springer Science+Business Media LLC 2017

research materials and resources, we can easily apply mutants and marker lines to our culture system [7], indicating its usefulness for analyzing vascular cell differentiation by genetic approaches. In this chapter, we describe the methods for our culture system using Arabidopsis leaf disks.

## 2 Materials

### 2.1 Preparation of Plant Materials

1. MS agar plates: Prepare growth medium by dissolving 2.2 g/L MS Basal Medium (Sigma), 10 g/L sucrose, and 0.5 g/L 2-morpholinoethanesulfonic acid monohydrate (MES) in Milli-Q water and adjusting the pH to 5.7 with KOH. Autoclave with 0.6% Bacto™ Agar (BD). Pour into each square grid plates (95 mm×95 mm) inside a laminar flow hood and let it dry to form MS agar plates (*see* **Note 1**). Store at 4 °C up to 2 months.
2. Sterilizing solution: Add 0.1 mL of 10% Triton-X100 in 1 mL of sodium hypochlorite solution and dilute to 10 mL with Milli-Q water. The solution should be prepared fresh.
3. Purified, autoclaved Milli-Q water.
4. Surgical tape.
5. Continuous light chamber (22 °C, 45–55 μmol/m$^2$/s).

### 2.2 Culture of Leaf Disks

1. Induction medium stock: Prepare 2.2 g/L MS Basal Medium containing 50 g/L d(+)-glucose (*see* **Note 2**) in Milli-Q water and adjust to pH 5.7 with KOH (*see* **Note 3**). Dispense in small glass bottles as appropriate and autoclave at 120 °C for 20 min to sterilize. The prepared medium can be stored at room temperature for a couple of weeks.
2. 2,4-D stock: Dissolve in ethanol and dilute with purified autoclaved Milli-Q water up to 2.5 g/L, and then sterilize using 0.22 μm filter. Aliquot 200 μL in sampling tubes and store them at −20 °C.
3. Kinetin stock: Dissolve and dilute with 0.1 M KOH up to 0.5 g/L, and then sterilize using 0.22 μm filter. Aliquot 200 μL in sampling tubes and store them at −20 °C.
4. Bikinin stock: Dissolve and dilute with DMSO up to 10 mM, and then sterilize using 0.22 μm filter (*see* **Note 4**). Aliquot 50 μL in sampling tubes and store them at −20 °C.
5. Sterilized 12-well plate.
6. Sterilized petri dish.
7. 1 mm disposable biopsy punch.
8. 1 μL Inoculating Loop.

9. Surgical forceps.
10. Continuous light chamber (22 °C, 60–70 μmol/m$^2$/s).
11. Rotary shaker.

### 2.3 Observation of Induced Xylem Cells

1. Fixative solution: Mix acetic acid and ethanol in the ratio of 1:3. Store at room temperature.
2. Clearing solution: Mix chloral hydrate, glycerol, and Milli-Q water in the ratio of 8:1:2 (w/v/v). Store at 4 °C.

## 3 Methods

### 3.1 Preparation of Plant Materials

1. Add 1 mL of sterilizing solution to the 1.5 mL sampling tube containing Arabidopsis seeds. Gently shake to mix and then use a rotator to stir for 5 min.
2. Spin down and transfer the tube inside a laminar flow hood. Allow to stand for 5 min.
3. Remove the supernatant using a P1000 pipette. Add 1 mL of purified, autoclaved Milli-Q water and mix well, and allow to stand until seeds sink to the bottom of the tube. Repeat this step four times.
4. Store the sterilized seeds at 4 °C for 2 days in order to keep the germination timing constant (*see* **Note 5**).
5. Sow 13 seeds per MS agar plate (*see* **Note 6**). Dry the plate inside a laminar flow hood for around 15 min. Seal the plate with surgical tape.
6. Incubate for 3–4 weeks under continuous light (22 °C, 45–55 μmol/m$^2$/s) (Fig. 1).

**Fig. 1** Prepared plant materials. Arabidopsis plants are grown on MS agar plates for 3–4 weeks under continuous light

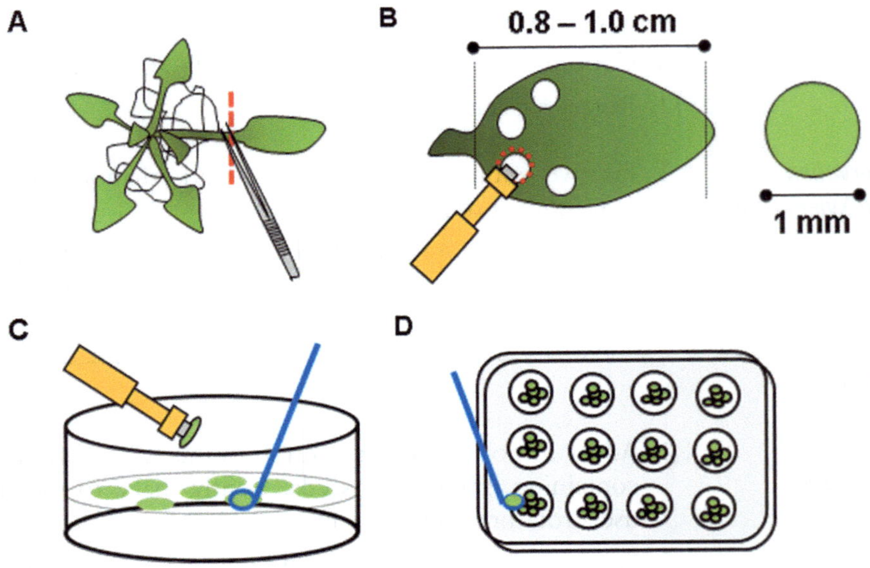


**Fig. 2** Schematic diagram of the preparation of leaf disks. (**a**) True leaves are cut using forceps. (**b**) Leaf disks are clipped out from the leaves using biopsy punch. (**c**) Leaf disks are pumped into a petri dish containing control medium. (**d**) Leaf disks are transferred to multi-well plate containing BIK medium using inoculating loop

### *3.2 Culture of Leaf Disks*

1. Defrost the stock of hormones (2,4-D and kinetin) and bikinin at room temperature before the start of the experiment. Transfer the tubes inside a laminar flow hood and add 50 μL of 2,4-D stock, 50 μL of kinetin stock to 100 mL of sterilized induction medium stock (control medium). Measure 50 mL of this medium and add 50 μL of bikinin stock (BIK medium) (*see* **Note 7**).
2. Add 2.5 mL of the prepared BIK medium into each well of a 12-well plate and pour some of the remaining control medium into a petri dish (*see* **Note 8**).
3. Cut Arabidopsis true leaves from 3–4-week-old plants across the petiole using forceps and select the ones with leaf blade length of 0.8–1.0 cm (*see* **Note 9**) (Fig. 2a). Wear double rubber gloves and place 3–4-week-old rosette leaves of Arabidopsis onto an index finger upside down (the adaxial side of the leaf faces to finger). Using a biopsy punch of 1 mm diameter on the gloved hand, carefully clip out leaf disks from the basal part of the rosette leaves (*see* **Note 10**) (Fig. 2b). Insert a P1000 pipette at the back of the biopsy punch and pump air into it to push out the disks into the petri dish containing control medium (*see* **Note 11**) (Fig. 2c).
4. Transfer 10–15 of the collected leaf disks to each well containing the BIK medium using inoculating loop (*see* **Note 11**) (Fig. 2d). Seal the 12-well plate with surgical tape.

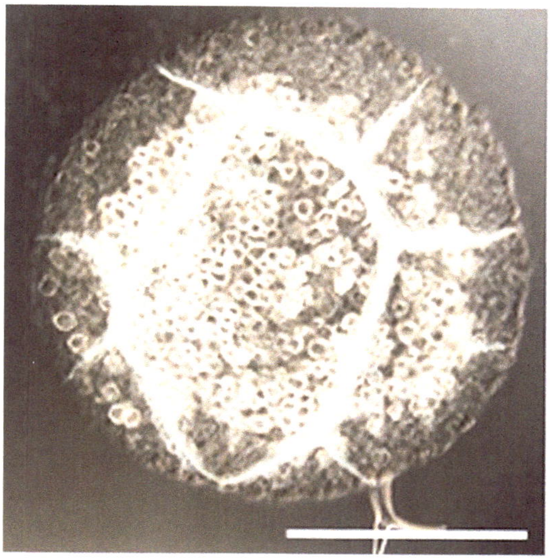

**Fig. 3** A leaf disk after 3 days of induction. Induced xylem cells with thickened cell walls can be distinguished from normal mesophyll cells. Differentiation rates in this culture system, defined as the ratio of xylem area to the whole leaf disk area, are approximately 50–70 %. Bar indicates 500 μm

5. Culture the leaf disks for 3 days under slightly stronger light conditions (60–70 μmol/m²/s) than during seedling growth with shaking at 110 rpm on the rotary shaker (*see* **Note 12**).

### 3.3 Observation of Induced Xylem Cells

1. Fix the cultured leaf disks in fixative solution at room temperature overnight.
2. Transfer into clearing solution and leave for over 3 h.
3. Mount on a glass slide with the clearing solution and observe under bright field microscope (Fig. 3).
4. Switch to UV exposure to observe autofluorescence from xylem secondary cell walls for more visible images.

## 4 Notes

1. No less than 40 mL of medium is added to 95 mm×95 mm plates to prevent desiccation of medium during plant growth.
2. Glucose is vital for induction of vascular differentiation and its concentration should be strictly maintained. Previous experiments using sucrose instead of glucose as a substrate in the induction medium stock resulted in inhibition of ectopic vascular differentiation.
3. The adjustment of pH is very critical when preparing the induction medium stock. The margin of error is strictly limited within 1 % (5.65–5.75) to ensure proper induction of vascular

differentiation. Due to the absence of MES buffer, it is difficult to adjust the pH of induction medium. We recommend to use 0.1 M KOH to avoid a sharp increase in pH of the induction medium stock.

4. Wear rubber gloves while handling bikinin for skin protection.
5. After washing with sterilizing solution, we store the seeds at 4 °C to align the germination stage of the seeds.
6. A maximum number of 13 seeds should be sowed with sufficient interspace on each solid MS agar plate. Growth at high density may disrupt proper leaf development due to stress.
7. Repeated freezing and thawing of hormones and bikinin reduces the efficiency of differentiation and is strictly not recommended. Therefore, it is useful to aliquot in specified volumes required during each experiment. Do not leave the reagents at room temperature for too long after defrosting. It may result in deterioration of their bioactivity. 2,4-D powder is prone to settle at the bottom of the tube after defrosting and should be mixed properly using a pipette to ensure complete dissolution before adding to the induction medium stock.
8. The volume of BIK medium in each well can be optimized according to the size of the well. We generally pour 1 mL in 24-well plates, 2.5 mL in 12-well plates, and 8 mL in 6-well plates. The volume of control medium in petri dish should be adequate to protect the leaf disks prepared in the next step from drying.
9. When cutting true leaves across the petiole, care should be taken not to harm the tissue using the sharp ends of the forceps. Close the lid of the MS agar plates after cutting true leaves to prevent desiccation. We have obtained the most efficient and stable vascular differentiation using leaf sizes between 0.8 and 1.0 cm [7].
10. Double rubber gloves are indispensable to avoid skin injury from the biopsy punch. We recommend clipping out leaf disks from the basal sides, because we have achieved faster differentiation rate with basal leaf disks compared to apical ones [7]. Also, we have specified the size of the leaf disks to 1 mm because 2–3 mm diameter disks are prone to curve, hence not suitable for observation.
11. The leaf disks are highly prone to desiccation and cannot be left in open air for long. Therefore, the leaf disks are floated on petri dishes containing control medium immediately after clipping. This step is also included to align the induction period of the leaf disks inside the BIK medium. After preparation, the leaf disks are collectively transferred into multi-well plates containing BIK medium. Inoculating loops are used instead of

forceps to prevent physical damage to the leaf disks. The number of disks per well can be optimized according to the well size to ensure maximum efficiency of vascular differentiation.

12. The light intensity for induction of vascular differentiation is stronger than in plant growth. Poor light can lead to reduced induction efficiency. Shaking the multi-well plates is not necessary, but recommended to improve aeration inside the well.

## Acknowledgment

This work was supported by Grants-in-Aid from the Ministry of Education, Culture, Sports, Science and Technology of Japan (15H01226 to Y.K., and 15H05958 and NC-CARP project to H.F.), from the Japan Society for the Promotion of Science (26891005 to Y.K. and 23227001 to H.F.) and from Naito Foundation to H.F.

### References

1. Fukuda H, Komamine A (1980) Establishment of an experimental system for the study of tracheary element differentiation from single cells isolated from the mesophyll of Zinnia elegans. Plant Physiol 65:57–60
2. Kubo M, Udagawa M, Nishikubo N, Horiguchi G, Yamaguchi M, Ito J, Mimura T, Fukuda H, Demura T (2005) Transcription switches for protoxylem and metaxylem vessel formation. Genes Dev 19:1850–1860
3. Oda Y, Iida Y, Kondo Y, Fukuda H (2010) Wood cell-wall structure requires local 2D-microtubule disassembly by a novel plasma membrane-anchored protein. Curr Biol 20: 1197–1202
4. Pesquet E, Korolev AV, Calder G, Lloyd CW (2010) The microtubule-associated protein AtMAP70-5 regulates secondary wall patterning in Arabidopsis wood cells. Curr Biol 20: 744–749
5. Kondo Y, Ito T, Nakagami H, Hirakawa Y, Saito M, Tamaki T, Shirasu K, Fukuda H (2014) Plant GSK3 proteins regulate xylem cell differentiation downstream of TDIF-TDR signalling. Nat Commun 5:3504
6. Kondo Y, Fukuda H (2015) The TDIF signaling network. Curr Opin Plant Biol 28:106–110
7. Kondo Y, Fujita T, Sugiyama M, Fukuda H (2015) A novel system for xylem cell differentiation in Arabidopsis thaliana. Mol Plant 8:612–621

# Chapter 6

# VND6-induced Xylem Cell Differentiation in Arabidopsis Cell Cultures

Yoshihisa Oda

## Abstract

In vitro xylem differentiation is a powerful technique that can be used to elucidate the process of xylem development that occurs deep inside plant tissues in nature. The experimental procedure described here is designed to induce metaxylem vessel differentiation at exceptionally high frequency and synchronicity using genetically engineered Arabidopsis cell suspensions. By triggering a transcriptional switch, over 80 % of the cells synchronously differentiate into xylem cells within 32 h of treatment with estradiol. Exogenous marker genes can be transiently introduced into the cells by coculturing them with transformed Agrobacterium before inducing xylem differentiation. This system is fast, easy to handle, and highly compatible with molecular and cell biology techniques used to explore xylem cell differentiation.

**Key words** Xylem, Tracheary element, VND6, Arabidopsis, Secondary cell wall

## 1 Introduction

Xylem cell differentiation is a dynamic process during which cellular function dramatically changes through active secondary cell wall deposition. In vitro xylem differentiation is a powerful technique for elucidating the cellular and molecular mechanisms underlying xylem development, which occur deep inside plant tissues in nature. A number of methods have been developed to induce xylem differentiation in vitro using a variety of materials. An experimental system involving the use of mechanically isolated zinnia mesophyll cells [1] has long been used to explore xylem and vascular development. During the past decade, experimental systems using *Arabidopsis thaliana* have been developed. These systems have enabled researchers to utilize the many Arabidopsis biological resources and techniques to extensively examine gene function during xylem development. These systems typically depend on the manipulation of phytohormones [2–6]. On the other hand, several systems have been used to achieve in vitro differentiation of specific xylem types, such as protoxylem, metaxylem, and fibers by

Miguel de Lucas and J. Peter Etchells (eds.), *Xylem: Methods and Protocols*, Methods in Molecular Biology, vol. 1544,
DOI 10.1007/978-1-4939-6722-3_6, © Springer Science+Business Media LLC 2017

activating the corresponding master transcription factors [7–9]. Here, I describe a method for metaxylem vessel differentiation, which can be induced at exceptionally high frequency and high synchronicity using a transgenic Arabidopsis suspension cell line [8]. This cell line was genetically engineered to harbor *VND6*, which encodes a master transcription factor for metaxylem vessel differentiation [5] under the control of an estradiol-inducible promoter [10]. Addition of estradiol to the culture medium immediately activates *VND6* expression, which in turn triggers differentiation into metaxylem vessel elements. Upon activation of *VND6*, over 80% of cells undergo differentiation within 32 h. I also show that exogenous marker genes can be transiently introduced into the cells by coculturing them with Agrobacterium before inducing xylem differentiation. This process enables rapid visualization of proteins of interest in living xylem cells. This culture system is rapid, easy to handle, highly efficient, and amenable for molecular and cell biology studies investigating xylem cell differentiation.

## 2 Materials

Prepare all solutions using ultrapure water or deionized water and analytical grade reagents. Prepare and store all reagents at room temperature (unless otherwise indicated). Carefully follow all waste disposal regulations when disposing of waste materials.

### 2.1 Medium and Regents for Weekly Subculture

1. 100 mL Erlenmeyer flask.
2. Silicon cap (Shin-Etsu Polymer Co., Ltd., cat. no. BIO-SILICO N-28).
3. Sterile pipette.
4. Sucrose.
5. Murashige and Skoog (MS) plant salt mixture (Wako Chemical Co., Ltd.; cat. no. 392-00591).
6. 2,4-dichlorophenoxyacetic acid (2,4-D) stock solution (1000×): 1 mg/mL 2,4-D in water. Store at 4 °C.
7. B5 vitamin stock solution (100×): 0.4 g/L nicotinic acid, 0.4 g/L pyridoxine-HCl, 4 g/L thiamine-HCl, and 40 g/L myo-inositol. Dissolve 40 mg nicotinic acid, 40 mg pyridoxine-HCl, 400 mg thiamine-HCl, and 4 g myo-inositol and adjust the volume to 100 mL with water. Divide the solution into 10 mL aliquots in 15 mL tubes. Store at −20 °C.
8. MS medium: dissolve 4.6 g MS plant salt mixture and 30 g sucrose in 800 mL of water. Add 1 mL of 2,4-D stock solution and 10 mL of B5 vitamin stock solution. Dilute to 1 L with water. Adjust to pH 5.8 with KOH. Divide the medium into 15 mL aliquots in 100 mL Erlenmeyer flasks (*see* **Note 1**).

**Fig. 1** Preparation of MS medium for cell culture. From left: double layer of aluminum foil, silicon cap, 100 mL flask capped with a silicon cap, 100 mL flask with a double layer of aluminum foil over the silicon cap

Cap the flasks with BIO-SILICO and cover the top of each flask with a double layer of aluminum foil (Fig. 1). Sterilize by autoclaving (120 °C, 20 min).

9. Hygromycin B stock solution (1000×): 50 mg/mL (final concentration, 50 mg/L) solution in water. Sterilize the solution through a 0.22 μm pore size syringe filter. Store at −20 °C.

### 2.2 Medium and Regents for Xylem Differentiation

1. 250 mL glass bottles.
2. Induction medium (2,4-D free MS medium): dissolve 4.6 g MS plant salt mixture and 30 g sucrose in 800 mL of water. Add 10 mL B5 vitamin stock. Dilute to 1 L with water. Adjust to pH 5.8 with KOH. Divide the medium into 200 mL aliquots in 250 mL glass bottles (*see* **Note 1**). Sterilize by autoclaving (120 °C, 20 min).
3. 17-β-estradiol stock solution (5000×): 10 mM (final 2 μM) solution in dimethyl sulfoxide (DMSO). Store at −20 °C.
4. Brassinolide stock solution (5000×): 10 mM (final concentration, 2 μM) solution in DMSO. Store at −20 °C.
5. Sterile 15 mL tube.
6. Sterile 12-well multiwell plate.

### 2.3 Observation of Xylem Cells

1. Alexa Fluor®594-conjugated wheat germ agglutinin (WGA, Thermo Fisher Scientific Inc.; cat. no. W11262, *see* **Note 2**) stock solution (1000×): 1 mg/mL solution (final concentration, 1 mg/L) in water. Store at −20 °C.
2. Glass slide.
3. Coverslip (22 mm × 22 mm).

### 2.4 Transformation

1. 250 mL glass bottles.
2. MS medium for transformation: dissolve 4.6 g MS plant salt mixture and 10 g sucrose in 800 mL of deionized water. Add 10 mL of B5 vitamin stock solution and 1 mL of 2,4-D stock solution. Adjust to 1 L with water. Adjust to pH 5.8 with KOH. Divide the medium into 200 mL aliquots in 250 mL glass bottles (*see* **Note 1**). Sterilize by autoclaving (120 °C, 20 min).
3. Claforan (Sanofi Aventis) stock solution (200×): 100 mg/mL (final concentration, 500 mg/L) solution in water. Sterilize through a 0.22 μm pore size syringe filter. Store at −20 °C.
4. 50 mL Erlenmeyer flask. Cover the top of the 50 mL Erlenmeyer flask with a double layer of aluminum foil. Sterilize by autoclaving (120 °C, 20 min).
5. Acetosyringone (3′,5′-dimethoxy-4′-hydroxyacetophenone, Sigma-Aldrich Co. LLC; cat. no. D134406-1G) stock solution (2000×): 100 mg/mL (final concentration, 50 mg/L) solution in DMSO. Store at −20 °C.

### 2.5 Equipment

1. Clean bench.
2. Rotary shaker.
3. Fluorescence microscope and/or confocal laser microscope.

## 3 Methods

All procedures should be performed on a clean bench.

### 3.1 Maintaining the Cell Line

1. Maintain the cell line [8] in liquid culture on a rotary shaker (set at 120 rpm) at 22 °C in the dark. Every 7 days, transfer 12 mL of culture (*see* **Note 3**) to 15 mL of fresh MS medium using a sterile pipette and add 15 μL of hygromycin stock solution.

### 3.2 Inducing Xylem Vessel Differentiation

1. Transfer a 0.5 mL aliquot of 7-day-old culture (*see* **Note 4**) to a sterile 15 mL tube.
2. Add 4.5 mL of induction medium, 1 μL of estradiol stock, and 1 μL of brassinolide stock.
3. Mix the culture gently and divide into 1 mL aliquots in a 12-well multiwell plate.
4. Culture the plate on a rotary shaker (set at 120 rpm) at 22 °C in the dark.

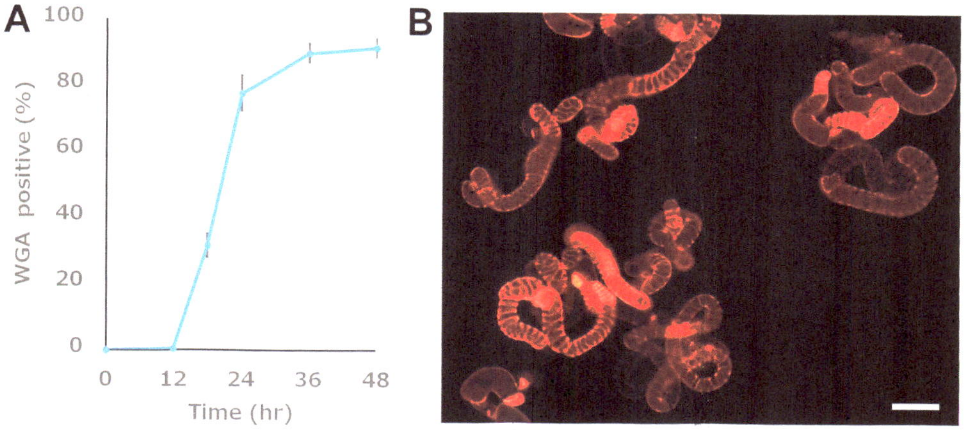


**Fig. 2** Typical image of xylem differentiation in cultures. (**a**) Time course showing the percentage of wheat germ agglutinin (WGA)-positive cells after the addition of estradiol. Data are expressed as the mean ± SD of three cultures (*n* > 300 cells). (**b**) Fluorescence microscope image showing xylem cells after 36 h of culture. Fluorescent signals indicate secondary cell walls stained with Alexa Fluor®594-conjugated WGA. Note the patterned secondary cell walls. Scale bar, 30 μm

### *3.3 Observing Differentiating Xylem Cells*

1. Add 1 μL of fluorescein-conjugated WGA stock solution to each well and culture for at least 10 min on a rotary shaker (*see* **Note 5**).
2. Place a 20 μL aliquot of the culture onto a glass slide, cover with a coverslip, and observe the specimen under a fluorescence microscope at 100× magnification. The secondary cell walls of differentiated xylem cells are stained with fluorescein-conjugated WGA (Fig. 2). Typically, secondary cell wall deposition begins approximately 18 h after the onset of culture. The frequency of WGA-positive cells should reach 80% during a 24–32 h period (*see* **Note 6**).

### *3.4 Transformation*

1. Culture Agrobacterium cells on solid LB medium.
2. Transfer a 2.5 mL aliquot of 7-day-old Arabidopsis cell culture (*see* **Note 7**) to a sterile petri dish and remove the medium using a micropipette (*see* **Note 8**).
3. Add 4 mL of MS medium and 2.5 μL of acetosyringone stock solution to the cells.
4. Suspend the cells gently in the medium and transfer the cell suspension to a sterilized 50 mL flask.
5. Transfer a small amount of fresh Agrobacterium cells directly from solid LB medium to the plant cell culture and suspend them using a 200 μL micropipette (*see* **Note 9**).
6. Cover the top of the flask with a double layer of aluminum foil. Culture the cells on a rotary shaker (set at 120 rpm) at 22 °C in the dark for 44 h.

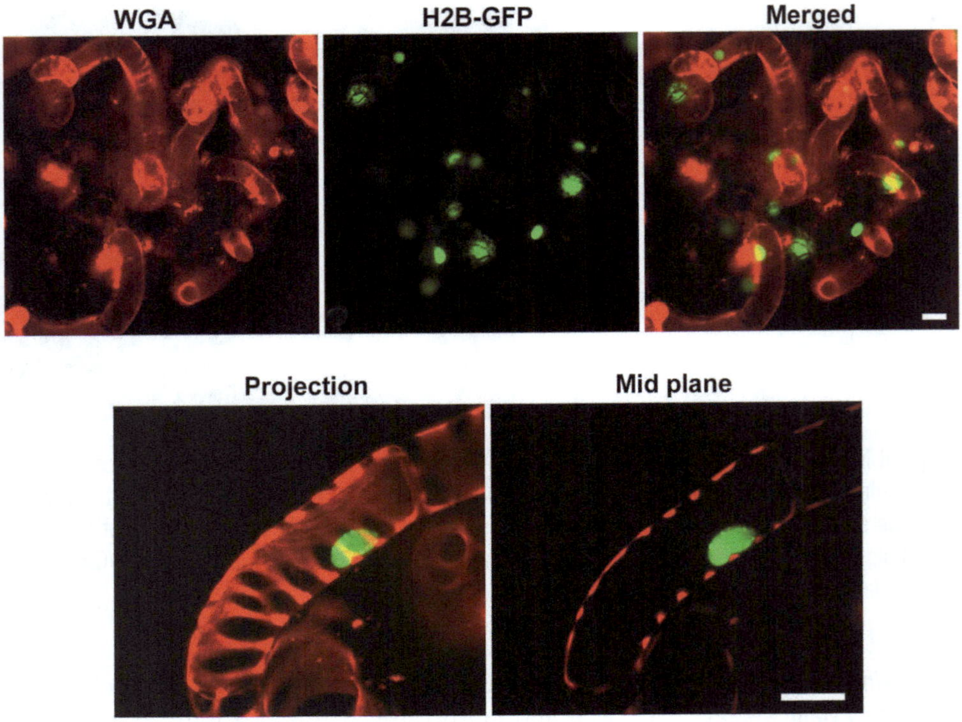


**Fig. 3** An example of cells transformed with *35S:histoneH2B-GFP*. The nuclei of differentiating xylem cells are labeled with H2B-GFP. (*Top*) Epi-fluorescence microscopy. (*Bottom*) Confocal microscopy, focused on a differentiating xylem cell. Scale bar, 20 μm

7. Add 5 mL of fresh MS medium and 50 μL of Claforan stock solution. Culture the cells on a rotary shaker for 4 days.
8. The culture can be used to induce xylem differentiation (Fig. 3) and/or for weekly subculture.

## 4 Notes

1. The pH-adjusted medium can be stored in polypropylene bottles at −20 °C. Avoid repeated freeze–thaw cycles.
2. As an alternative, Alexa Fluor®488-conjugated WGA (Thermo Fisher Scientific Inc., cat. no. W11261) can be used.
3. Depending on the speed of cell proliferation, the volume of cell culture to transfer should be between 10 and 15 mL.
4. 6–9-day-old cells should be used for efficient xylem differentiation.
5. WGA does not affect the progression of xylem differentiation.
6. The efficiency of xylem differentiation largely depends on the condition of the cells that are used. Keep a stationary weekly subculture and use fully proliferated, fresh cells.

7. 7–9-day-old cells are suitable for transformation.
8. To avoid sucking up the cells when removing the medium; use a 200 μL micropipette equipped with a fine-pointed tip.
9. Use newly cultivated, fresh Agrobacterium for high transformation efficiency. Typically, 1–10 % of the cells are transformed. The transformation efficiency largely depends on the gene used for transformation.

## Acknowledgment

This work was supported by grants from MEXT KAKENHI (project no. 16H01247 and 15H01243), the JSPS KAKENHI (project no. 16H06172), and the Mitsubishi Foundation (to Y.O.)

### References

1. Fukuda H, Komamine A (1980) Establishment of an experimental system for the study of tracheary element differentiation from single cells isolated from the mesophyll of Zinnia elegans. Plant Physiol 65:57–60
2. Oda Y, Mimura T, Hasezawa S (2005) Regulation of secondary cell wall development by cortical microtubules during tracheary element differentiation in Arabidopsis cell suspensions. Plant Physiol 137:1027–1036
3. Sawa S, Demura T, Horiguchi G, Kubo M, Fukuda H (2005) The ATE genes are responsible for repression of transdifferentiation into xylem cells in Arabidopsis. Plant Physiol 137: 141–148
4. Kondo Y, Fujita T, Sugiyama M, Fukuda H (2015) A novel system for xylem cell differentiation in Arabidopsis thaliana. Mol Plant 8: 612–621
5. Kubo M, Udagawa M, Nishikubo N, Horiguchi G, Yamaguchi M, Ito J et al (2005) Transcription switches for protoxylem and metaxylem vessel formation. Genes Dev 19:1855–1860
6. Pesquet E, Korolev AV, Calder G, Lloyd CW (2010) The microtubule-associated protein AtMAP70-5 regulates secondary wall patterning in Arabidopsis wood cells. Curr Biol 20: 744–749
7. Ohashi-Ito K, Oda Y, Fukuda H (2010) Arabidopsis VASCULAR-RELATED NAC-DOMAIN6 directly regulates the genes that govern programmed cell death and secondary wall formation during xylem differentiation. Plant Cell 22:3461–3473
8. Oda Y, Iida Y, Kondo Y, Fukuda H (2010) Wood cell-wall structure requires local 2D-microtubule disassembly by a novel plasma membrane-anchored protein. Curr Biol 20: 1197–1202
9. Yamaguchi M, Goue N, Igarashi H, Ohtani M, Nakano Y, Mortimer JC et al (2010) VASCULAR-RELATED NAC-DOMAIN6 and VASCULAR-RELATED NAC-DOMAIN7 effectively induce transdifferentiation into xylem vessel elements under control of an induction system. Plant Physiol 153: 906–914
10. Zuo J, Niu QW, Chua NH (2000) Technical advance: an estrogen receptor-based transactivator XVE mediates highly inducible gene expression in transgenic plants. Plant J 24: 265–273

# Part II

## Imaging

# Chapter 7

# Live Imaging of Developing Xylem In Planta

Raymond Wightman

## Abstract

The xylem vessel develops from a long cylindrical cell that deposits a patterned secondary wall and ending with programmed cell death. The dynamic arrangement of cell wall enzymes, membranes and the cytoskeleton can be recorded using live fluorescence imaging. The protocol presented here focuses upon the microscopy of intracellular components in developing vessels of the root using either epifluorescent or confocal microscopes.

**Key words** Microscopy, Confocal, Cytoskeleton, Cellulose synthase, Organelles, Roots, Arabidopsis

## 1 Introduction

The developing xylem vessels present some challenges for live imaging, given their accessibility deep within the tissue and the deposition of lignified secondary walls. Visualization of structures and processes within the cells relies heavily on fluorescent constructs. For this reason most xylem imaging has been carried out in *Arabidopsis*. When designing fluorescent constructs, it is important to consider what amount of detail needs to be recorded. Constructs containing the commonly used *35S* promoter can exhibit poor expression within the stele region of the root. Other more ubiquitously expressed reporters may be observed in the developing vessels but signal coming from the same compartments in neighboring cells can both make identification of the vessels difficult (they only deposit secondary walls at a late stage prior to cell death) and mask weaker signals at or close to the cortex. Furthermore, the depth required for imaging may make very small fluorescent structures look diffuse. These technical challenges can be overcome by making use of xylem (trans)differentiation of epidermal cells or in culture (*see* Chapters 4–6 in this book); however, the observation of the long cylindrical cells that become functional water conduits can only be carried out in planta, the focus of this chapter.

Miguel de Lucas and J. Peter Etchells (eds.), *Xylem: Methods and Protocols*, Methods in Molecular Biology, vol. 1544,
DOI 10.1007/978-1-4939-6722-3_7, © Springer Science+Business Media LLC 2017

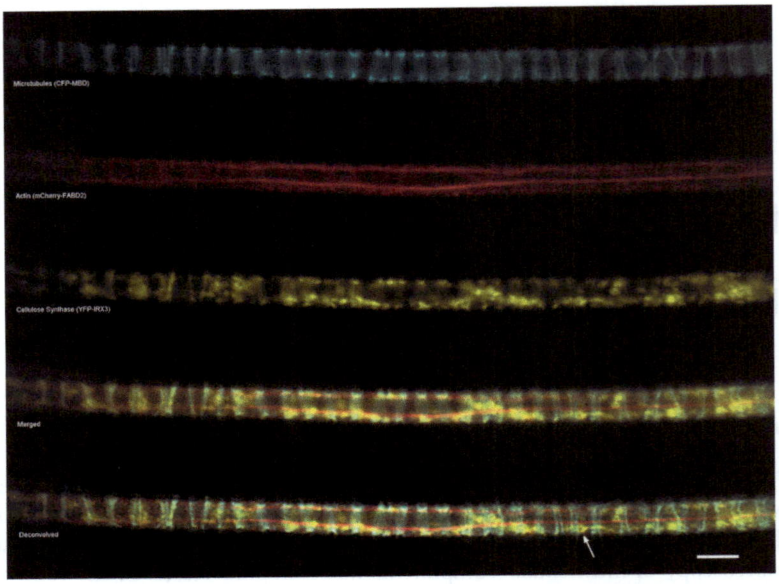


**Fig. 1** Confocal images showing microtubules (*cyan*, CFP-MBD), actin (*red*, mCherry-FABD2), and cellulose synthase (*yellow*, YFP-IRX3) in the same developing xylem vessel. A merged image before and after deconvolution is shown. Note the small improvement in clarity of ring-shaped Golgi labeling by YFP-IRX3 (*arrow*) in the deconvolved data. Images were acquired using a Leica SP8 confocal equipped with a 63× 1.4 NA oil objective and HyD detectors. Line averaging = 8, pinhole = 1.7 airy units. Deconvolution was carried out using Huygens software. Scale bar = 5 μm

The secondary walls of xylem contain mostly cellulose, synthesized by a complex comprising three CESA subunits [1]. One of these subunits encoded by *IRX3/CESA7*, when fused to YFP, can be clearly seen in the developing xylem, but not neighboring cells. The *IRX3* promoter has been used to express reporters of the cytoskeleton and endomembranes specifically in developing xylem [2, 3]. Images from a selection of these reporters are shown in Fig. 1. Microtubules bundle and form beneath future secondary walls and colocalize with the thick bands of YFP-IRX3. Actin is observed as bundles running along the long axis of the cell that give rise to transverse fibers at intervals close to wall thickenings. This cytoskeletal network gives rise to fast transport highways as observed by the Golgi labeled with either tagged IRX3 or mannosidase I [3]. In addition to cellulose, secondary walls also contain the hemicellulose, xylan, and imaging has been extended to the cognate biosynthetic enzymes [4] (Paul Dupree, personal communication).

## 2 Materials

### 2.1 Plant Materials and Growth Conditions

1. *Arabidopsis* seeds are surfaced sterilized using a mixture of sodium hypochlorite (30% v/v) and Triton X-100 (1% v/v) for 8–10 min, rinsed with sterile water three to five times, and stratified for 3 days at 4 °C.
2. Murashige and Skoog media: 0.5× Murashige and Skoog salts containing B5 vitamins, 15 g/L agar.
3. Stratified seeds are plated and placed vertically at 22 °C in the light for 5–6 days.
4. *Optional*: Reagents for pharmacological manipulation, e.g., Oryzalin (Sigma-Aldrich).

### 2.2 Microscopy Equipment

1. Any epifluorescence or confocal microscope equipped with either a water immersion NA 1.2 objective or an oil immersion objective with NA greater than 1.3. It is preferable to have a lens with a coverslip correction collar.
2. For epifluorescence microscopes a monochrome CCD or EMCCD camera.
3. Slides, coverslips, tissue, nail varnish (optional).
4. *Optional*: Software for deconvolution and 3D rendering.
5. *Optional*: Color camera and dual YFP/mCherry filter cube (Chroma Technology Corp part no.: 51019).

## 3 Methods

### 3.1 Sample Mounting

While I have had some limited success with imaging late-developing xylem from longitudinal cuts of young stem tissue, routine imaging and treatments make use of intact roots placed on a microscope slide.

1. Using forceps and held by the cotyledons, up to four seedlings are carefully removed from the agar and placed side by side on a glass slide with the roots lying completely flat.
2. Drops of water (or other aqueous immersion medium) are immediately applied to the roots to prevent drying.
3. A coverslip is applied taking care to cover the roots but not the green parts of the seedling. The resulting sample should have the roots beneath the coverslip and all the hypocotyls exposed along one edge.
4. Excess water is removed by gently touching the edge of the coverslip with tissue. Be careful not to apply any pressure to the coverslip.

5. For immediate imaging slides can be placed on the microscope stage, for time-lapse imaging slides can be sealed along three sides with nail varnish.

### 3.2 Epifluorescence Microscopy

1. Find the root tip using brightfield and then observe reflected light from the sample using the filter cube that best corresponds to the fluorescent protein under observation.
2. Track back from the root and one or two developing vessels should be emitting signal. Focus the objective on one vessel.
3. If the fluorescence signal is blurry despite focus adjustments then use a longer working objective or, if present, turn the correction collar to allow greater depth. For example, a Leica HCX PL APO CS 63× 1.2 NA water objective should have the collar set to above 0.18.
4. For taking images, the user needs to carefully balance exposure time to limit photobleaching to camera gain, which should be set as low as possible without losing signal from the vessels.
5. Where camera gain is high, better signal-to-noise ratio can be obtained by taking two images in quick succession followed by averaging. Some camera software has this function, otherwise take two images manually and average using freely available software such as imageJ.
6. Where fast moving objects appear blurred, reduce the exposure time by one third and set camera binning to 2 × 2.
7. *Optional*: For multichannel imaging of more than one fast moving component (e.g., Golgi) a dual filter cube coupled with a sensitive color camera means that both channels are acquired concurrently. The fluorescent protein tags need to be spectrally distinct. I have found the YFP/mCherry dual filter set to be suitable for fast time-lapse in epifluorescence systems [3].

### 3.3 Confocal Microscopy (Point Scanning)

1. Using the eyepieces, find the root tip using brightfield and then observe reflected light from the sample as for epifluorescence above. Focus the objective on one vessel.
2. Select your light settings based on the fluorescent proteins you want to observe. Where possible, use a 440 nm laser for CFP (and derivatives) rather than a 458 nm line. Where available use a high sensitivity detector (*see* **Note 1**).
3. For excessively noisy (grainy) images, increase the pinhole diameter and lower the detector gain. Typically the pinhole is set to 1.7 airy units for sensitive confocals (*see* **Note 2**).
4. Set the line averaging as needed, typically set to 4 or above (8 or above for a resonant scanner).

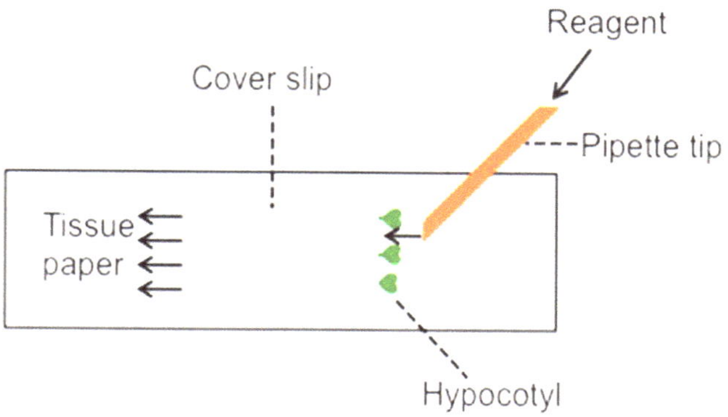


**Fig. 2** Diagram showing drug treatment method for mounted seedlings

5. For more efficient scanning, rotate the scan field so that the vessels are oriented horizontally and use a letterbox format (e.g., 1024 × 256).
6. If fast moving organelles appear blurred then increase the scan speed and/or reduce the amount of averaging (*see* **Note 3**).

### 3.4 Drug Treatments

1. For drug treatments, the coverslip should not be sealed. Take image acquisition 1 (no treatment) control and keep the slide on the stage.
2. Without moving the sample carefully add the drug in 5 μl amounts to the edge of the coverslip lined with hypocotyls, use a micropipette for this. Meanwhile place a thin piece of tissue to the opposite edge of the coverslip. Water will be seen to be drawn in to the tissue (Fig. 2).
3. Maintain addition of the drug to the edge of the coverslip. Change wet tissue for dry as necessary. For a 25 × 25 mm coverslip, at least 40 μl of drug should be flushed through the coverslip.
4. Take acquisitions every 1–5 min.
5. Effect of drugs such as Oryzalin (microtubule depolymerization) and Lantrunculin B (actin depolymerization and stationary organelles) will be observed within minutes [3].

### 3.5 Observing Onset of Programmed Cell Death (PCD)

1. Take roots containing at least two xylem-specific reporters with a pH sensitive fluorescent protein (e.g., YFP-IRX3) and a pH tolerant reporter (mCherry-ER) (*see* **Note 4**).
2. Take time-course at 1–2 min intervals of developing vessels containing thick secondary walls (observed using brightfield).

3. The beginning of PCD is observed as a sudden decrease in YFP signal that occurs in a "wave". mCherry signal is unaffected [5].
4. YFP signal is eliminated leaving just the mCherry signal and is additionally associated with changes in endomembrane morphology.

## 4 Notes

1. To limit unwanted autofluorescence, a 514 nm laser or YFP excitation filter should be used for both GFP and YFP observations rather than shorter wavelengths. For vessels with thicker, more developed secondary walls, CFP and GFP are best avoided and fluorescent proteins with longer wavelengths (e.g., YFP and mCherry) should be used.
2. 3D rendering is best achieved on confocal datasets with a pinhole diameter set close to 1 airy unit or deconvolved. A small (<0.4 μm) step size should be set for z-acquisitions.
3. For confocal imaging, an increase in clarity of organelle structures can be obtained with deconvolution software using either an experimental or theoretical point spread function (Fig. 1). This is useful where the pinhole diameter has been increased to greater than 1.5 airy units.
4. For imaging of newly forming vessels, a promoter other than from *IRX3* will be required. For ubiquitous expressing reporters, the precise file of cells giving rise to protoxylem can be observed by viewing one channel containing a xylem specific reporter such as YFP-IRX3.

### References


1. Taylor NG, Howells RM, Huttly AK et al (2003) Interactions among three distinct CesA proteins essential for cellulose synthesis. Proc Natl Acad Sci U S A 100:1450–1455
2. Wightman R, Marshall R, Turner SR (2009) A cellulose synthase-containing compartment moves rapidly beneath sites of secondary wall synthesis. Plant Cell Physiol 50:584–594
3. Wightman R, Turner SR (2008) The roles of the cytoskeleton during cellulose deposition at the secondary cell wall. Plant J 54:794–805
4. Brown D, Wightman R, Zhang Z et al (2011) Arabidopsis genes IRREGULAR XYLEM (IRX15) and IRX15L encode DUF579-containing proteins that are essential for normal xylan deposition in the secondary cell wall. Plant J 66:401–413
5. Young B, Wightman R, Blanvillain R et al (2010) pH-sensitivity of YFP provides an intracellular indicator of programmed cell death. Plant Methods 6:27

# Chapter 8

# Immunolocalization in Secondary Xylem of *Populus*


## Suzanne Gerttula and Andrew Groover



## Abstract

Immunolocalization can be used to precisely visualize the location of specific proteins, cell wall components, or any other molecules within cells or tissues for which specific antibodies are available. Here we describe an immunolocalization protocol for tissue sections of woody *Populus* stems. The protocol includes descriptions of the required sectioning, fixation, probing, detection, and imaging parameters, as well suggested controls useful in interpreting results.



**Key words** Antibodies, Antigen, Cell wall, Forest trees, Immunolocalization, Wood development


## 1 Introduction

Immunolocalization has been a primary tool in animal and plant research. There are numerous variations on the technique, but in general immunolocalization makes use of antibodies to detect and image the location of specific molecules within in a tissue (Fig. 1). In plants, a wide range of antibodies have been raised against individual moieties contained within proteins, carbohydrates, chromatin marks, and even small molecules including hormones. To probe and detect these epitopes in plant tissues, typically the tissue is fixed, embedded and sectioned. The sections are affixed to slides and probed with the antibody (termed the primary antibody). Alternatively, sections or small pieces of tissue can be probed as "whole mounts" without being affixed to slides, as in the protocol here. Two different methods can be used to then visualize where the primary antibody has localized the antigen. In colorimetric methods, a secondary antibody recognizing the primary antibody is conjugated to an enzyme (e.g., alkaline phosphatase) and used to probe the section. A substrate is then provided which, when acted upon by the conjugated enzyme, produces a colored reaction product that can be visualized under the microscope. An alternative method is to use a fluorescent reporter for visualization. In this approach, a fluorophore is conjugated to the secondary antibody,


Miguel de Lucas and J. Peter Etchells (eds.), *Xylem: Methods and Protocols*, Methods in Molecular Biology, vol. 1544,
DOI 10.1007/978-1-4939-6722-3_8,

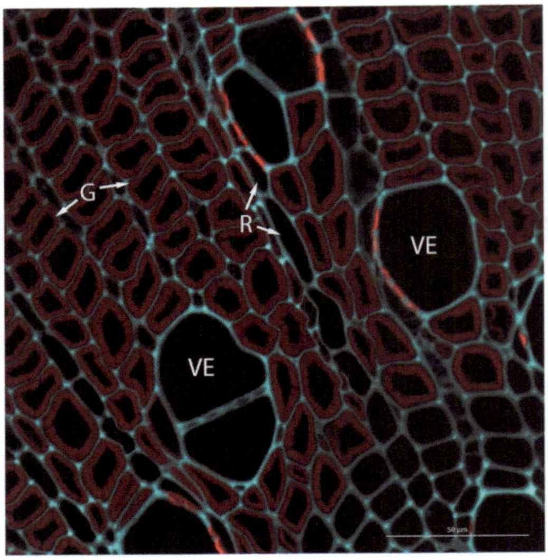


**Fig. 1** Confocal image of immunolocalization in secondary xylem of *Populus*. Vibratome sections were prepared from *Populus* stems of a genotype overexpressing the ARBORKNOX2 transcription factor, and subjected to the immunolocalization protocol described here. The primary antibody used was JIM14 at 1:500 dilution (Complex Carbohydrate Research Center, University of Georgia, Athens, GA), whose recognition includes arabinogalactan protein epitopes [1]. Secondary antibody was biotin-xx-goat anti-rat IgG at 1:400 dilution (Life Technologies A10517). Detection was with 4 μl/ml of Streptavidin Alexa Flour 488 (Life Technologies S11223). *Red signal* is from Alexa Flour 488, which includes G-layers of tension wood fibers. *Blue signal* is autofluorescence resulting from UV excitation. Imaging was performed using a Zeiss LSM 710 confocal microscope. *G* g-layer, *R* ray, *VE* vessel element

which can be visualized using a confocal or epifluorescence microscope. A modification of this approach is to conjugate biotin to the secondary antibody, and perform detection using a fluorophore conjugated to streptavidin, which has high affinity to biotin. This method essentially serves to amplify the observed signal, and is utilized in the protocol here (Fig. 1).

While relatively simple in concept, immunolocalization must be optimized for different species, tissues, antibodies and epitopes. Xylem tissues from woody plants are challenging subjects. When sectioned, xylem has a tendency to curl and it can be difficult to firmly adhere sections to slides. Additionally xylem contains a number of secondary compounds and has cells with complex, lignified cell walls. These features frequently cause problems including nonspecific binding of antibodies, high background autofluorescence in fluorescence-based detection and difficulties in reaching the target epitope. Thus, the protocol here has been optimized to reduce nonspecific binding in xylem using complex blocking reagents and copious washing steps. Additionally, the protocol makes use of a fluorophore for detection that has excitation and emission spectra that have acceptable levels of autofluorescence in xylem. Finally, a whole mount procedure is used to avoid problems with adherence of sections to slides, and the sectioning and fixation procedures optimize perfusion of paraformaldehyde through the complex woody cell wall thus enhancing antibody access to the target epitope.

The immunolocalization protocol here makes use of whole-mount sections of stem tissues from *Populus* seedlings, and has been used previously to visualize both specific protein and cell wall epitopes [2]. The same protocol could likely be extended to xylem of other woody species.

## 2 Materials

Having a high quality primary antibody is critical for success. If working with an untested antibody, it is recommended that specificity be tested using Western blots. Note that formaldehyde is a carcinogen and should be handled using appropriate personal protective equipment within a fume hood. Formaldehyde must be disposed of as hazardous waste following disposal regulations.

### 2.1 Sectioning and Fixation

1. Freshly prepared FAA fixative on ice: 11.25 ml EtOH, 10 ml $dH_2O$, 1.25 ml acetic acid, 2.5 ml formalin (BP531 Fisher).
2. PBT buffer: 5 ml PBS, 45 ml $dH_2O$, 100 μl Tween 20.
3. 1× blocking buffer: 2.5 ml 10× PBS, 5 ml 10% Fish Gelatin, 2.5 ml 10% BSA, 15 ml $dH_2O$, 50 μl Tween 20.
4. Required equipment: Vibratome, double edged razor blades, fine tipped paintbrushes, deionized water, cyanoacrylate adhesive (e.g., Crazy Glue™), lab quake or similar mixer, 5 ml round bottom tubes, 1.5 ml snap top tubes.

### 2.2 Probing, Detection and Imaging

1. PBT buffer: 5 ml PBS, 45 ml $dH_2O$, 100 μl Tween 20.
2. 1× blocking buffer: 2.5 ml 10× PBS, 5 ml 10% Fish Gelatin, 2.5 ml 10% BSA, 15 ml $dH_2O$, 50 μl Tween 20.
3. Primary antibody.
4. Secondary antibody that recognizes primary antibody, conjugated to biotin.
5. Strepavidin conjugated Alexa 488 (ThermoFisher S11223).
6. Required equipment: Labquake or similar mixer, microscope slides and coverslips, epiflourescence or confocal microscope.

## 3 Methods

### *3.1 Sectioning and Fixation*

1. Place the vibratome directly adjacent to a fume hood. Add water to the vibratome and attach the blade as per manufacturer instructions. The water level should cover the cutting edge of the razor blade. Note that the blade angle, speed of blade advance, and amplitude of blade can all be adjusted to optimize sectioning.
2. Prepare FAA fixative and place it in 2–3 ml round bottom 5 ml tubes on ice inside the fume hood.
3. From a healthy tissue culture grown poplar seedling (or similar plant), use a razor blade to cut a segment from the stem to be sectioned. The segment should be no more than ~6 mm long. For difficult to section woody stems (e.g., soil grown poplar seedlings, or those over 5 mm in diameter), cut longitudinally into two half-rounds (*see* **Note 1**).
4. Affix the stem segment to the sectioning block by placing a drop of cyanoacrylate adhesive (e.g., super glue) on the block by dipping the base of the stem segment in the glue, and then positioning the segment vertically near one edge of the block. Avoid placing the stem segment in a drop of glue that has been placed on that block as this can cause ineffecient glue drying. Allow the glue to dry for ~2 min before placing the sectioning block into the chuck of the vibratome.
5. Adjust the sample height so that the blade is just above the height of the sample.
6. Use the control switches on the vibratome to control the blade movement. Up will fast forward, down will reverse, and up/release will advanse the blade slowly to cut.
7. Press up on the control switch to move the blade towards the sample, then use slow advance as the blade cuts the section. Before reversing the blade, adjust the sample height down by 20–30 μm to avoid scraping the sample as the blade returns to the starting position.
8. Raise the block by 50–100 μm and take a fresh cut to leave a smooth surface.
9. After making the first full section, use a small paintbrush to retrieve it from the water bath and transfer it to fixative on ice.
10. The blade must be sharp, and should be changed every few sections or every section for woodier tissue.
11. Once you have collected as many sections as desired, remove the round bottom tube from ice and place it into a vacuum chamber at room temperature in the fume hood. Draw a vacuum and incubate for 45 min (*see* **Note 2**).

12. Repeat until all tissue types have been collected. Note, if there are two halves of the same stem that both need to be cut, they can both be placed onto the cutting block and sectioned at the same time. The blade dulls very easily and cutting half rounds helps to prevent that. Using a fresh blade for every cut helps. If the sample bends as it is cut, a convex or incomplete section results and several passes of the remaining tissue are sometimes required before another full section can be obtained. Smaller samples can be embedded in agarose before being affixed to the block for sectioning.

### 3.2 Blocking and Probing

1. After fixing for 45 min, remove the samples from the vacuum and place them on ice. Step the samples slowly into PBT as follows. Over the course of ~5 min, add PBT slowly by drops to equal the volume of fixative. Remove 1 ml of the solution and repeat slowly adding drops of PBT for a total of 1 ml. This process is repeated several times until the schlieren lines are no longer apparent.
2. Remove the remaining fixative/PBT solution and replace it with 2 ml of PBT. Cap the tube and incubate it with gentle mixing 15 min at room temperature.
3. Discard the PBT wash solution, add 2 ml of PBT and incubate at room temperature with gentle mixing for 15 min. Repeat for a total of 4, 15 min washes. Distribute the sections into Eppendorf tubes. (Note: Include at minimum one treatment for a no-primary antibody control, and one or more treatments for the primary antibody. If the primary antibody has not been previously optimized in your system, consider testing a range of primary antibody concentrations.)
4. Remove the PBT and replace it with a small volume of blocking buffer to rinse. Replace with 1 ml of blocking buffer and incubate at room temperature for ≥1 h.

### 3.3 Primary Antibody Incubation

1. Primary antibodies are typically shipped and stored at high concentration and need to be diluted to effective working concentrations by the user. A good rule of thumb is to start out with a manufacturer's recommended final working concentration and then try one sample at tenfold higher concentration, and one sample at tenfold lower concentration if you have enough antibody on hand (*see* **Notes 3–5**). Briefly centrifuge the primary antibody in a microcentrifuge before using to avoid pipetting any aggregates.
2. Place the Eppendorf tubes with sample and blocking buffer on ice. Pipette the required volume of primary antibody into each tube, with the exception of the no-primary antibody control (*see* below). Place the samples on a lab quake or similar rocker and mix gently at 4° overnight.

3. The next day, remove the tubes from the 4° chamber. Remove the block/antibody solution and perform one PBT wash. Then perform a series of four PBT washes of 15 min each, at room temperature with gentle mixing.
4. Remove the PBT and replace it with a small volume of blocking buffer to rinse. Replace with 1 ml of blocking buffer and incubate at room temperature for ≥1 h.

### 3.4 Secondary Antibody Incubation

1. The secondary antibody should recognize antibodies from the species used to produce the primary antibody. For example, if the primary antibody was produced in rabbits, the secondary antibody should recognize rabbit IGG (sometimes termed an anti-rabbit secondary antibody).
2. Secondary antibodies are typically purchased in a highly concentrated form and need to be diluted to effective working concentrations. Follow the manufacturers recomendations or test a dilution series of concentrations.
3. The secondary antibody used in this protocol should be labeled with biotin.
4. Briefy spin the secondary antibody in a microcentrifuge, and pipette the desired volume into your tube containing sample and blocking buffer, on ice.
5. Incubate overnight at 4° C with gentle mixing.

### 3.5 Dectection and Imaging

1. Rinse your samples one time with PBT, then perform four PBT washes, 15 min each at room temperature with gentle mixing.
2. Rinse one time with a small volume of blocking buffer, then replace with 1 ml of fresh blocking buffer and incubate at room temperature with gentle mixing for >30 min.
3. Add streptavidin-fluorophore conjugate to desired concentrations. In our lab, we use 4 μl/ml streptavidin-Alexa 488. Wrap each tube in foil to protect from direct light, to avoid photobleaching of the fluorophore.
4. Incubate at room temperature for ≥30 min with gentle mixing.
5. Rinse one time with 1 ml PBT.
6. Wash four times with 1 ml PBT, 15 min each at room temperature with gentle mixing.
7. Sections are now ready to be imaged by confocal or epifluorescence microscopy.

## 4 Notes

1. A potential area requiring optimization is the sectioning of xylem tissue. The best settings for vibratome sectioning must be experimentally determined for specific samples and tissues. In general, the vibratome is limited by the thickness and density of what it can section, so if a sample is too large or hard to section, partial dissection may be beneficial (e.g., instead of sectioning the entire circumference of a stem, bisect it longitudinally and section half the stem). Alternatively, samples that are not rigid enough can be embedded in agarose before being sectioned.
2. Fixation can be optimized, although in this case the whole mount sections are relatively uniform and cells are relatively accessible to fixative. If the reader is working with a readily transformed species and needs to determine protein localization, an alternative approach is to visualize recombinant, fluorescently labeled proteins in whole mount sections [3]. Consulting other protocols for plant immunolocalization may be useful when adapting or optimizing your protocol for other sample types [4].
3. The most unpredictable part of immunolocalization using this protocol will be the primary antibody. General information about working with and characterizing antibodies is available [5]. If you are working with a new antibody, it is highly recommended that you test a wide range of concentrations, and include a well-characterized antibody as a positive control in your initial experiments.
4. Several on-line resources are available for finding antibodies of interest. For example, a wide range of antibodies recognizing cell wall epitopes is available through the Complex Carbohydrate Research Center at the University of Georgia (http://www.ccrc.uga.edu/~mao/wallmab/Antibodies/antib.htm). Examples of immunolocalization for both protein and cell wall epitopes can be seen in [2].
5. Additional experimental validation of a new antibody such as western blotting can also helpful in determining specificity and estimating titer. In all experiments, a "no-primary antibody" treatment should be included, to determine the extent and nature of background fluorescence from autofluorescence or nonspecific sticking of the secondary antibody or streptavidin-fluorophore conjugate. In general, washes can be extended in time or number without significant loss of signal.

## Acknowledgments

We thank Julie Lee for assistance with confocal microscopy, and the UC Davis Department of Plant Biology for access to confocal microscopy facilities. This protocol was developed using funds from USDA AFRI grant 2015-67013-22891. Any use of trade, product, or firm names is for descriptive purposes only and does not imply endorsement by the US Government.

### References

1. Knox JP et al (1991) Developmentally regulated epitopes of cell surface arabinogalactan proteins and their relation to root tissue pattern formation. Plant J 1(3):317–326
2. Gerttula S et al (2015) Transcriptional and hormonal regulation of gravitropism of woody stems in populus. Plant Cell 27(10): 2800–2813
3. Sauer M et al (2006) Immunocytochemical techniques for whole-mount in situ protein localization in plants. Nat Protoc 1(1):98–103
4. Pasternak T et al (2015) Protocol: an improved and universal procedure for whole-mount immunolocalization in plants. Plant Methods 11:50
5. Howard G, Kaser M (eds) (2006) Making and using antibodies. CRC Press, Boca Raton, FL

# Chapter 9

# Monitoring Vascular Regeneration and Xylem Connectivity in *Arabidopsis thaliana*

**Charles W. Melnyk**

## Abstract

Plants have a remarkable ability to regenerate vascular tissue after damage or wounding. A striking example of this phenomenon is the cutting and rejoining of plants during the process of grafting, which humans have used for millennia. Here, I describe how to graft *Arabidopsis* seedlings and how to monitor the vascular reconnection process during wound healing using fluorescent dyes and fluorescent proteins. Furthermore, I describe how to visualize xylem formation during graft healing. These techniques are useful for studying how the vascular system regenerates and for better understanding the process of plant grafting.

**Key words** *Arabidopsis thaliana*, Micro-grafting, Chimeric plants, Fluorescent dyes, Vascular regeneration, Xylem formation

## 1 Introduction

The plant vascular system has an incredible ability to regenerate and reform. The mechanism for healing remains largely unknown, but this system presumably exists to restore the systemic transport of water, nutrients and signaling molecules after damage by herbivory, pathogens, or mechanical wounding. The field of vascular regeneration is growing, due in part to an interest in how the vascular system functions, how tissues and cells regenerate, and how plants graft. People have grafted plants for millennia and this technique involves the cutting and joining of different plants to each other to propagate or improve the properties of the scion (grafted shoot) by grafting to rootstocks (grafted roots) with enhanced disease resistance or dwarfing abilities [1].

Previous work studying vascular regeneration used cutting or needle stabs to interrupt plant vasculature [2–5]. These studies often used vascular-mobile dyes including the fluorescent dye fluorescein to measure phloem connectivity [3]. For instance, the fluorescein derivative carboxyfluorescein diacetate (CFDA) can

Miguel de Lucas and J. Peter Etchells (eds.), *Xylem: Methods and Protocols*, Methods in Molecular Biology, vol. 1544, DOI 10.1007/978-1-4939-6722-3_9, © Springer Science+Business Media LLC 2017

be applied to a leaf to monitor phloem transport from source to sink [6] or applied to the root to monitor xylem movement from root to shoot [7]. Phloem transport can also be monitored by expressing the Green Fluorescent Protein (GFP) in the phloem companion cells and observing GFP movement into non-transgenic grafted tissue [7, 8]. Other methods to study vascular regeneration have included sectioning through the wound or graft junction to look at the differentiation of xylem elements and their connectivity at the wound site [2, 4, 5, 9]. Many of these studies were performed in non-model species such as *Coleus*, *Zinnia*, or *Pisum*, or used large tissues such as inflorescence stems, making genetic analyses or imaging difficult.

Here, I use *Arabidopsis* hypocotyl grafting in young seedlings as a model for studying vascular regeneration. Grafting in *Arabidopsis* was first described over 20 years ago [10] and is routinely used to generate chimeric organisms to study the non-cell autonomous actions of RNAs, proteins and hormones [1]. Multiple tissues of *Arabidopsis* are suitable for grafting, including inflorescence stems [11], cotyledons [12], shoot apical meristem with developing leaves [13], mature shoot/mature root [14], and the young shoot/young root [15]. The most routine procedure is grafting young *Arabidopsis* seedlings between the shoot and root since many plants can be rapidly and efficiently assembled.

The method presented here is a fast and effective way to graft young *Arabidopsis* seedlings that with practice gives a success rate of over 80% with up to 80 grafts assembled per hour with wild-type plants. By grafting a wild-type scion expressing GFP in the phloem companion cells (*pSUC2::GFP*) to a non-transgene rootstock, grafted plants can be monitored daily for phloem reconnection by the appearance of GFP in the rootstock vasculature. Using such an assay, the genetic requirement of the rootstock can be easily assayed since root genotypes can be varied. GFP movement from rootstock to scion is not observed so testing the genetic requirements in the scion requires the genotype of interest to be crossed into the *pSUC2::GFP* background. I also present a method whereby grafts are assayed for vascular connectivity by applying CFDA to the rootstock to test for xylem connectivity or applying CFDA to the scion to test for phloem connectivity [7]. These assays allow almost any genotype to be tested for vascular regeneration. Finally, I present a method to visualize xylem formation at the graft junction based on the method of Malamy and Benfey [16]. With successful grafting, xylem differentiation begins above the graft junction after 5 days and at 7 days, the majority of wild-type plants have formed connecting xylem elements above and below the graft junction.

## 2 Materials

### 2.1 Arabidopsis Grafting

1. Sterile *Arabidopsis* seed.
2. Agar plates suitable for growing *Arabidopsis*, such as ½ Murashige and Skoog (MS) media, pH 5.7, 1 % agar.
3. Whatman 3MM Chr filter paper, 46 × 57 cm.
4. Hybond N membrane, 20 cm × 3 m.
5. Aluminum foil.
6. Ultra Fine Micro Knives (Fine Science Tools catalog number 10315-12; *see* **Note 1**).
7. Fine forceps.
8. 9 cm round petri dishes.
9. 70 % ethanol.
10. Sterile, autoclaved deionized water.
11. Portable dissecting microscope.
12. Laminar flow hood.
13. Parafilm.
14. 20–22 °C growth cabinet.

### 2.2 Fluorescein Movement Assays

1. Grafted plants (from Subheading 3.1).
2. 5(6)-Carboxyfluorescein diacetate (CFDA) in powdered form.
3. Dimethyl sulfoxide (DMSO).
4. Fine forceps.
5. Fluorescent dissecting microscope with a GFP or YFP filter.
6. Plant agar with a higher pH, such as ½ MS with 0.8 % agar at pH 6.7.
7. 9 cm round Petri dishes.

### 2.3 Tissue Clearing for Xylem Visualization

1. Grafted plants (from Subheading 3.1).
2. 2 ml microcentrifuge tubes (*see* **Note 2**).
3. Six well cell culture plates (*see* **Note 2**).
4. Cell strainers for 50 ml tubes (*see* **Note 2**).
5. Hydrochloric acid.
6. Sodium hydroxide.
7. Ethanol.
8. Methanol.
9. Microscope slides and cover slips.
10. Microscope with DIC optics.

## 3 Methods

### 3.1 *Arabidopsis* Hypocotyl Grafting and Phloem Connectivity Assays

1. Place sterilized *Arabidopsis* seed on solid media such as ½ Murashige and Skoog (MS) medium containing 1 % agar. Use media without selective antibiotics or herbicides, as this can negatively affect plant growth. For phloem connectivity assays, include *pSUC2::GFP* expressing seeds in the Col [17] or Ler [18] background. Seal the plates and leave in the dark at 4 °C for 2–7 days to stratify seeds.
2. Place plates vertically in a 20–22 °C growth chamber with 80–100 μmol/m²/s of light to ensure hypocotyls and roots grow in a straight line. Under short day conditions (8 h light), plants are grafted 7 days after moving out of stratification conditions. Under long day conditions (16 h light), plants are grafted 5 days after moving out. I typically use short day grown plants as the hypocotyls are longer and easier to graft.
3. Cut the Hybond membrane in 2.5 × 4 cm rectangles, and the Whatman paper in ~8.5 cm circles that will fit inside a 9 cm petri dish. Two Whatman circles and one Hybond rectangle are required per grafting plate. Cut also rectangles of Whatman paper (~3 × 8 cm) that will be used during grafting to add and remove water. Wrap the Hybond rectangles and Whatman paper in aluminum foil and autoclave.
4. Perform all grafting and subsequent steps in a sterile laminar flow hood using a stereomicroscope that has been sterilized by wiping with 70 % ethanol. Grafting tools (forceps and micro knives) are sterilized by placing in 70 % ethanol and allowing to dry thoroughly before grafting. The grafting tools are periodically placed in 70 % ethanol to re-sterilize during the grafting procedure. Letting the ethanol evaporate after dipping is important as ethanol in the grafting plate will inhibit graft formation.
5. Two sterile Whatman circles are placed in a petri dish filled with sterile water. One Hybond rectangle is placed on top. Remove the three water soaked papers, let the excess water drip off, and place them in a new petri dish so that one Hybond rectangle is on top of two Whatman circles. During graft healing, the Hybond ensures newly formed roots don't anchor to the Whatman, whereas the Whatman maintains the correct moisture levels.
6. When plants are 5 or 7 days old (*see* **step 2**), place 6–12 *Arabidopsis* wild-type or mutant seedlings in a row onto the Hybond membrane (Fig. 1). Choose plants that have cotyledons at right angles to the hypocotyl, and hypocotyls that are straight and lying flat on the agar surface. When moving seedlings to the Hybond rectangle, take care not to damage

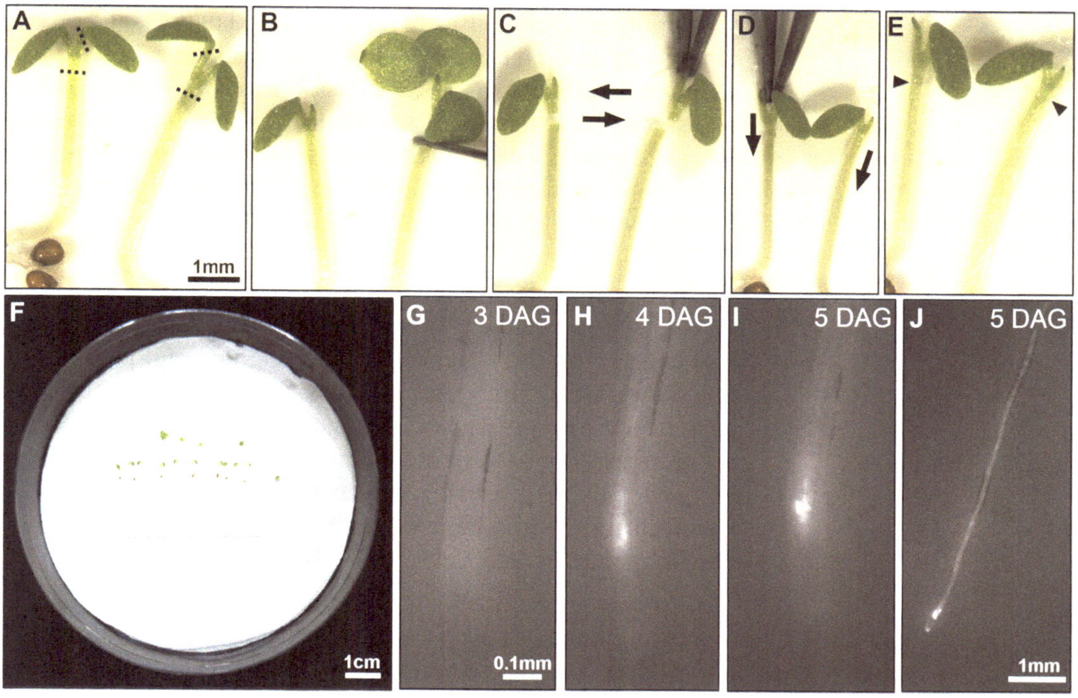


**Fig. 1** *Arabidopsis* hypocotyl grafting and GFP mobility assays. Two Whatman circles and one Hybond rectangle are placed in a petri dish with *Arabidopsis* seedlings on top. Seedlings are cut (*dotted lines*), scions moved to the desired rootstock (*arrows*), and then pushed into place (*arrows*) (**a**–**f**). The graft junction is denoted by triangles. *pSUC2::GFP* scions grafted to non-transgenic rootstocks initially show no fluorescence in the roots (**g**), but 3–4 days after grafting (DAG), show fluorescence in the root tip and vasculature (**h**–**j**) indicative of phloem reconnection

roots, hypocotyl or cotyledons. If grafting different genotypes to each other, place a second row of plants above the first row or alternate genotypes in the same row. For instance, *pSUC2::GFP*-expressing plants as even numbers and the wild-type or mutant rootstock as odd numbers.

7. Using the micro knife, cut off and remove one cotyledon from the scion, normally the cotyledon that is damaged, smaller or bent (Fig. 1a). Leaving one cotyledon allows the scion to lie flat and align more easily. Make a butt-end cut on the scion hypocotyl close to where the cotyledons emerge (a transverse cut 90° to the axis of elongation) (Fig. 1b). Make a clean cut and avoid crushing the tissue. Ensure there is some water visible on the Hybond membrane as this will make cutting easier. Discard the cut scion cotyledon and root. Make a second butt-end cut in the desired rootstock and discard the shoot. If the hypocotyl is not lying flat against the Hybond, gently roll the hypocotyl by pushing at the root–hypocotyl junction with closed forceps. Be careful not to damage the hypocotyl by gripping or crushing, as this will inhibit grafting.

8. Pick up the cut scion, usually by the exposed petiole, and place it close to the desired rootstock (Fig. 1c). For phloem connectivity assays, this will be a *pSUC2::GFP* scion to a wild type or mutant rootstock. For other assays (*see* Protocol 3.2 or 3.3), this could be a wild-type scion to a wild-type rootstock, mutant to mutant, or mutant to wild-type. Gently push on the scion meristem region, cotyledon, or exposed petiole with closed forceps until the cut scion comes into contact with the cut rootstock (Fig. 1d). Aligning the two tissues and gently pushing them together is sufficient for successful grafting (Fig. 1e). There should be some visible water on the plates to facilitate aligning and joining, but less water than during cutting. Excess water will make the scion stick to the forceps, whereas insufficient water will result in the scion not sticking to the root and becoming wilted. Sterilized rectangles of Whatman paper can be used to remove or add water. Plates can also be left open to dry in the flow hood for a couple minutes to reduce water levels.
9. Repeat this procedure for all the plants on the plate (Fig. 1f). If grafting becomes progressively harder as the plate dries out, add a drop or two of water to facilitate additional grafting. Once grafting is finished, water levels should be lower than for cutting and for graft alignment. Water should be barely visible as a slight shimmer around the hypocotyls and roots that lie perfectly flat. Not all hypocotyls will show this shimmer. Excess water is a major cause of graft failure and adventitious root formation. Once the water levels are correct, place the lid on the petri dish and seal with Parafilm. Mount the plates vertically in the growth chamber at 20–22 °C in 80–100 μmol/ $m^2$/s of light.
10. For phloem connectivity assays, plates should be checked at 48 h after grafting for movement of GFP from the *pSUC2::GFP* scion to the rootstock, and every 24 h after that until 7–10 days after grafting. GFP normally first appears in wild-type rootstocks 72 h after grafting most strongly in the root tips and to a lesser extent in the root vasculature (Fig. 1g–j). With some plants, fluorescence is initially subtle and gets stronger with time (Fig. 1g–i). In addition to monitoring fluorescent movement, rootstock growth can also be assayed by counting when plants resume primary root growth or form lateral roots, usually 4 or 5 days after grafting in wild-type Col-0 plants.
11. After 7–10 days on Whatman plates, successful grafts can be moved into soil or onto agar plates. Grafting success is measured by an absence of adventitious root formation above the graft junction, and the resumption of growth by the rootstock. With the above method, grafting success rates should approach 90% with experience.

### 3.2 CFDA Assays to Measure Xylem and Phloem Reconnection

1. Graft plants as in Protocol 3.1 but include an equal number of ungrafted controls that are treated identically to grafted plants, but not cut through the hypocotyl. Ungrafted controls can be included on the same plate as grafted plants using a second Hybond rectangle, or placed on a separate plate.
2. Phloem and xylem connectivity assays are performed 2–10 days after grafting when scion and rootstock are attached. Assays can be performed daily, in which case plants are discarded after the assay (since CFDA can persist in tissue for days) and new plants used for each time point. For single time point phloem assays, I use 4 days after grafting to assess whether a mutant is perturbed in phloem reconnection compared to wild type plants. For single time point xylem assays, I use 7 days after grafting to assess whether a mutant is perturbed in xylem reconnection compared to wild type plants. At these time points, over 80% of wild type Col-0 plants have phloem (4 days) or xylem (7 days) connections [7] so a deviation from these percentages indicates the mutation of interest may affect vascular reconnection.
3. For both xylem and phloem connectivity assays, dissolve 1 mg of CFDA per 10 μl of DMSO. Add 9.2 μl of this solution per 2 ml of ½ MS media, pH 6.7, 0.8% bacto agar that is molten (1 mM CFDA final concentration). The higher pH medium may improve solubility and fluorescence. In my experience, the solubility of CFDA depends on the manufacturer of the CFDA. If the CFDA does not dissolve completely at this concentration, reduce the amount of CFDA-DMSO solution added to the media.
4. For phloem assays, open the grafting plates and pipette a small drop (1–2 μl) of molten CFDA-agar onto the spot where previously, each cotyledon has been wounded by gently pressing on it with a fine forceps (Fig. 2a). Repeat for all grafted and ungrafted plants. One hour after CFDA-agar application, score the number of plants with fluorescence in the root vasculature below the root–hypocotyl junction using a fluorescent dissecting scope fitted with a GFP or YFP filter. Plants with functional phloem will have a clear and strong fluorescence signal in the roots (Fig. 2d and e). Between 90 and 100% of ungrafted controls should have a similar, or stronger, signal.
5. For xylem assays, pipette two lines each of 600 μl of molten CFDA-agar onto a 9 cm petri dish (one line parallel to the other). Let solidify and replace the petri dish lid to prevent the agar from drying out (Fig. 2).
6. For xylem assays, cut off the roots 1–3 mm below the hypocotyl–root junction of grafted plants and ungrafted controls using the micro knife. Using fine forceps, pick up the plant at

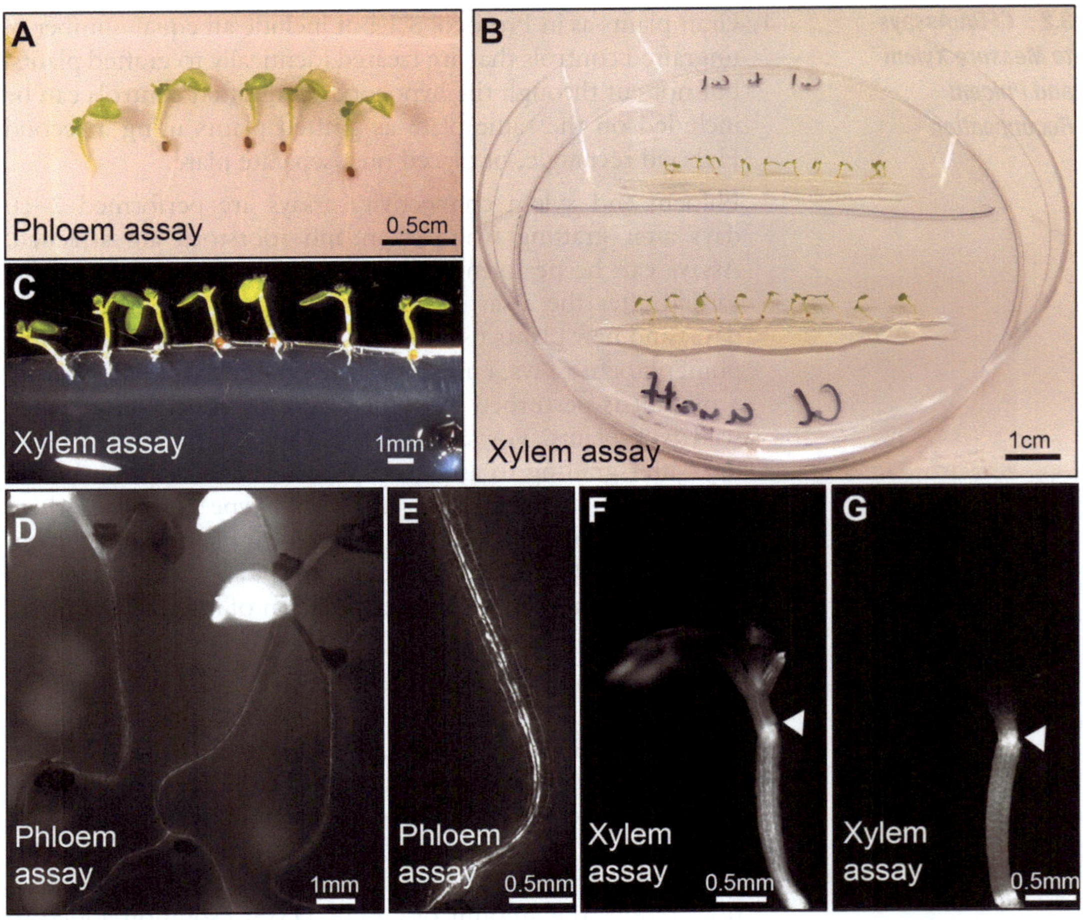


**Fig. 2** Phloem and xylem reconnection assays with carboxyfluorescein diacetate (CFDA). For phloem assays, cotyledons are gently wounded after grafting and a drop of 1 mM CFDA in agar is applied to the wound site (**a**). One hour after CFDA addition, the presence or absence of fluorescence in the rootstock is monitored as an indication of phloem connectivity (**d**, **e**). For xylem assays, the grafted seedlings are cut at the hypocotyl/root junction and the junction placed in a solution of 1 mM solidified CFDA-agar in a petri dish (**b**, **c**). 20 min after CFDA addition, the presence (**f**) or absence (**g**) of fluorescence in the scion is monitored as an indication of xylem connectivity, reprinted from [7] with permission from Elsevier. The graft junction is denoted by a triangle

the cut site and place the plant into the solidified CFDA-agar. Make sure the root–hypocotyl junction is in the agar, but that the graft junction and cotyledon do not touch the CFDA-agar (Fig. 2b and c). Grafted plants are placed in one CFDA-agar line, ungrafted controls in the second parallel line. Place the lid on the petri dish after all plants are moved.

7. For xylem assays, 20 min after CFDA-agar application, score the number of plants with fluorescence in the cotyledon or leaf vasculature using a fluorescent dissecting scope fitted with a GFP or YFP filter. Plants with functional xylem will have a clear

and strong fluorescence signal in the cotyledons and/or leaves (Fig. 2f and g). Between 90 and 100% of ungrafted controls should have a similar, or stronger, signal.

8. Once plants have been scored in the xylem assay, they can be immediately be used in the clearing assay (Subheading 3.3) to visualize xylem element formation.

### 3.3 Clearing of the Seedling to Visualize Xylem Formation

1. With successful grafting, xylem differentiation begins above the graft junction after 5 days. By 7 days, the majority of wild-type plants have formed xylem elements above and below the graft junction. To visualize xylem formation at the graft junction, the protocol by Malamy and Benfey [16] is modified slightly. Plants can be cleared at any age, but in my assays, I use 7 days after grafting as a standard time point to assess whether a mutant is perturbed in xylem formation compared to wild type plants. At 7 days, the majority of wild-type Col-0 plants have xylem differentiation above and below the graft junction [7].
2. Graft plants as in Protocol 3.1. At 7 days after grafting, cut off the *Arabidopsis* root with a razor blade or micro knife. Place 0.5–1 ml of 0.24 N HCl prepared in 20% methanol in a 2 ml microcentrifuge tubes and add the grafted plants to the solution. Many plants (>20) of the same genotype or graft combination can be added per tube, including grafted and ungrafted controls. Incubate at 57 °C for 30 min in a dry block heater. Avoid squeezing the graft junction and instead handle the plants via the hypocotyl–root junction or by the cotyledon. For a large number of genotypes (*see* **Note 2**), six well cell culture plates with cell strainers can be used instead of microcentrifuge tubes. In which case, float the culture plates in a water bath at 57 °C for 30 min.
3. Replace the solution with 7% NaOH in 60% ethanol and incubate at room temperature for 30 min. Take care when removing the HCl/methanol as the plants are fragile and will stick to the pipette tip.
4. Plants are rehydrated in subsequent washes of 40%, 20% and 10% ethanol for 5–10 min each.
5. Infiltrate the plants in 5% ethanol + 25% glycerol for 20 min. Plants can be stored in this solution for up to a month at 4 °C if necessary.
6. Place the samples in 50% glycerol and then transfer them to a microscope slide. If grafted and ungrafted controls are in the same tube, the presence or absence of the graft junction distinguishes them. I maintain the same orientation for all the samples on the slide and place them in rows to facilitate imaging. Afterwards, a cover slip is added.

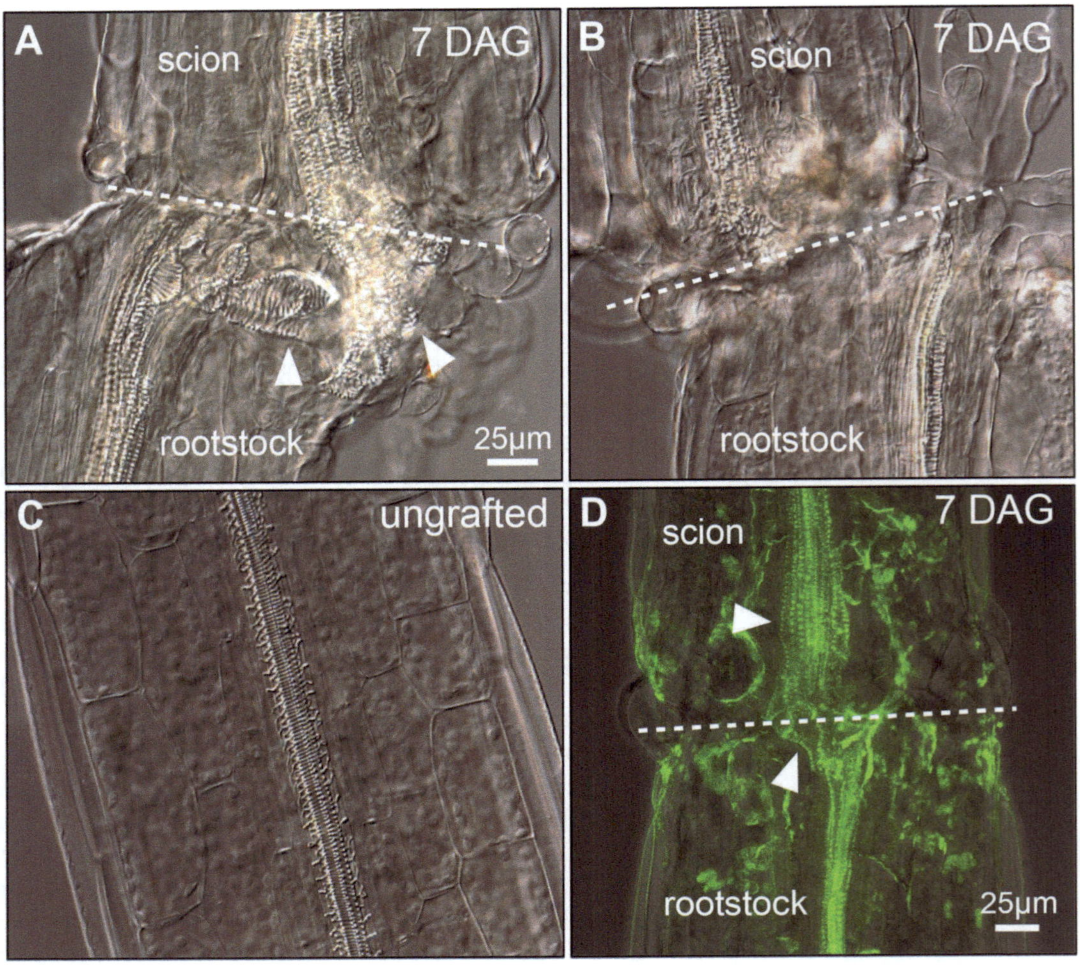


**Fig. 3** Xylem clearing assays to monitor xylem element formation and vascular continuity. After clearing the tissue, hypocotyls are mounted on slides and imaged using a microscope with DIC optics for new xylem formation that is continuous (**a**) or discontinuous (**b**) compared to ungrafted hypocotyls (**c**). A confocal microscope using GFP emission and excitation wavelengths can also detect new xylem element formation and continuity at the graft junction (**d**). Here, hypocotyls are imaged 7 days after grafting (DAG). Triangles denote newly differentiated continuous xylem elements

7. Xylem elements are visualized using a microscope with DIC optics (Fig. 3a–c). Alternatively, a light microscope with a GFP filter or confocal microscope with GFP settings (488 nm excitation, 500–600 nm emission) can detect xylem elements and Casparian strips (Fig. 3d) [19].

## 4 Notes

1. Grafting with Ultra Fine Micro Knives greatly improves grafting success, but double sided razor blades (split in half) will also work but with lower success rates. Ultra Fine Micro Knife

tips are fragile so care must be taken not to bend or damage them. The knives need to be replaced after several hundreds grafts or when cutting becomes less efficient.

2. For a large number (i.e., >12) of samples or genotypes that need to be cleared, using many microcentrifuge tubes is time consuming. Placing the samples in six well cell culture plates speeds up the process considerably. Initially, place one cell strainer per well, fill the well with solution (~5–10 ml) and add the desired samples. For solution changes, fill a new six well cell culture plate with the desired solution, and move the cell strainer containing the samples to the new plate.

## Acknowledgments

I thank Elliot Meyerowitz for critical reading. This work was funded through Gatsby Charitable Trust grants GAT3272/C and GAT3273-PR1 and by a Clare College Junior Research Fellowship.

### References

1. Melnyk CW (2016) Plant grafting: Insights into tissue regeneration. Regeneration. doi:10.1002/reg2.71
2. Flaishman MA, Loginovsky K, Lev-Yadun S (2003) Regenerative xylem in inflorescence stems of Arabidopsis thaliana. J Plant Growth Regul 22:253–258
3. Schulz A (1987) Sieve-element differentiation and fluoresceine translocation in wound-phloem of pea roots after complete severance of the stele. Planta 170:289–299
4. Nishitani C, Demura T, Fukuda H (2002) Analysis of early processes in wound-induced vascular regeneration using TED3 and ZeHB3 as molecular markers. Plant Cell Physiol 43:79–90
5. Jacobs WP (1952) The role of auxin in differentiation of xylem around a wound. Am J Bot 39:301–309
6. Oparka KJ, Duckett CM, Prior DAM, Fisher DB (1994) Real-time imaging of phloem unloading in the root-tip of Arabidopsis. Plant J 6:759–766
7. Melnyk CW, Schuster C, Leyser O, Meyerowitz EM (2015) A developmental framework for graft formation and vascular reconnection in Arabidopsis thaliana. Curr Biol 25:1306–1318
8. Yin H, Yan B, Sun J, Jia P, Zhang Z, Yan X, Chai J, Ren Z, Zheng G, Liu H (2012) Graft-union development: a delicate process that involves cell-cell communication between scion and stock for local auxin accumulation. J Exp Bot 63:4219–4232
9. Parkinson M, Yeoman MM (1982) Graft formation in cultured, explanted internodes. New Phytol 91:711–719
10. Rhee SY, Somerville CR (1995) Flat-surface grafting in Arabidopsis thaliana. Plant Mol Biol Report 13:118–123
11. Nisar N, Verma S, Pogson BJ, Cazzonelli CI (2012) Inflorescence stem grafting made easy in Arabidopsis. Plant Methods 8:50
12. Yoo SJ, Hong SM, Jung HS, Ahn JH (2013) The cotyledons produce sufficient FT protein to induce flowering: evidence from cotyledon micrografting in Arabidopsis. Plant Cell Physiol 54:119–128
13. Huang NC, Yu TS (2015) A pin-fasten grafting method provides a non-sterile and highly efficient method for grafting Arabidopsis at diverse developmental stages. Plant Methods 11:38
14. Chen A, Komives EA, Schroeder JI (2006) An improved grafting technique for mature Arabidopsis plants demonstrates long-distance shoot-to-root transport of phytochelatins in Arabidopsis. Plant Physiol 141:108–120
15. Turnbull CG, Booker JP, Leyser HM (2002) Micrografting techniques for testing long-distance signalling in Arabidopsis. Plant J 32:255–262
16. Malamy JE, Benfey PN (1997) Organization and cell differentiation in lateral roots of Arabidopsis thaliana. Development 124:33–44

17. Imlau A, Truernit E, Sauer N (1999) Cell-to-cell and long-distance trafficking of the green fluorescent protein in the phloem and symplastic unloading of the protein into sink tissues. Plant Cell 11:309–322

18. An H, Roussot C, Suarez-Lopez P, Corbesier L, Vincent C, Pineiro M, Hepworth S, Mouradov A, Justin S, Turnbull C et al (2004) CONSTANS acts in the phloem to regulate a systemic signal that induces photoperiodic flowering of Arabidopsis. Development 131:3615–3626

19. Alassimone J, Naseer S, Geldner N (2010) A developmental framework for endodermal differentiation and polarity. Proc Natl Acad Sci U S A 107:5214–5219

# Chapter 10

# Vascular Morphodynamics During Secondary Growth

**Pierre Barbier de Reuille and Laura Ragni**

## Abstract

Quantification of vascular morphodynamics during secondary growth has been hampered by the scale of the process. Even in the tiny model plant Arabidopsis thaliana, the xylem can include more than 2000 cells in a single cross section, rendering manual counting impractical. Moreover, due to its deep location, xylem is an inaccessible tissue, limiting live imaging. A novel method to visualize and measure secondary growth progression has been proposed: "the Quantitative Histology" approach. This method is based on a detailed anatomical atlas, and image segmentation coupled with machine learning to automatically extract cell shapes and identify cell type. Here we present a new version of this approach, with a user-friendly interface implemented in the open source software LithoGraphX.

**Key words** Quantitative histology, Cell segmentation, Arabidopsis hypocotyl, LithoGraphX

## 1 Introduction

Secondary growth generally occurs in seed plants, with the exception of monocotyledons [1, 2], and results in an increase in girth of plant organs. During this process, thousands of cells are produced by the vascular cambium, even in small, herbaceous plants. Key factors involved in regulating secondary growth have been identified in Arabidopsis thaliana (the current dicotyledon plant model), and several among them have also been shown to be important in tree species [3–7]. Model organisms have been generally selected for their small genome and their ability to grow easily in laboratory conditions due to their small size and short generation period [8]. Despite these characteristics, even in the tiny herbaceous Arabidopsis model plant, the xylem is too deep to be easily accessible using non-invasive methods such as live confocal microscopy (during secondary growth, it can be located more than 1 mm deep). Furthermore in adult plants it comprises a large number of cells, rendering manual quantification of the xylem morphodynamics impractical [9]. Thus in the literature, there are few quantitative studies of the xylem and most of them use holistic measures of secondary growth such as organ diameter, length or total area (i.e. of the stem or the root)

Miguel de Lucas and J. Peter Etchells (eds.), *Xylem: Methods and Protocols*, Methods in Molecular Biology, vol. 1544, DOI 10.1007/978-1-4939-6722-3_10, © Springer Science+Business Media LLC 2017

and xylem area [1, 9–12]. These holistic measures only coarsely describe secondary growth and do not catch the complexity of the process at the morphological level. Moreover, they can be misleading: Columbia (Col-0) and Landsberg (Ler), two Arabidopsis strains, differ in hypocotyl diameter, but while Col-0 is the thicker one, Ler has a larger xylem [9, 11].

Sankar et al. proposed a new approach to overcome these barriers and measure cell morphodynamics: the so called "Quantitative Histology approach" [9]. It combines a classical histological method to obtain cross sections of fixed samples, with innovative image analysis for data extraction and machine learning to automate cell type identification. It is based on four steps. (1) An anatomical atlas of organ cross sections at different developmental stages is built. (2) High resolution images of specific areas, such as the xylem, or of whole cross sections are acquired. (3) Cell shapes and positions are extracted through a pipeline of image preprocessing, image segmentation and features extraction. (4) A cell classifier is generated using machine learning methods to automate the process of cell identification. This approach has been successfully applied to the Arabidopsis hypocotyl in which seven types of cells, grouped in five classes, were successfully labeled: xylem vessels and parenchyma, xylem fibers, cambium, phloem bundles, phloem parenchyma and periderm, but it can be easily extended to other plant organs (root and stem) and/or other species. The quantification of the morphodynamics of the Arabidopsis hypocotyl lead to some new unexpected findings, such as that the cambium produces more phloem cells than xylem cells, highlighting the importance and the relevance of this approach for the study of secondary growth [9, 10]. A very similar approach has been recently developed to quantify laser scanning confocal sections by Hall et al. [13].

Here we present a modified protocol for the Quantitative Histology approach, with a user-friendly interface that is easier to install and execute by biologists. LithoGraphX, an open source software for the visualization and processing of biological images, was chosen for the implementation. Based on MorphoGraphX [13], it offers an environment with tools to manipulate images either through algorithms or direct user interactions. In addition, LithoGraphX offers an extension mechanism through which new image processing algorithms can be added in the form of libraries written in C++ or Python. Allowing extensions to be written in Python opens the door to a wide range of libraries written in this language, in particular for machine learning. Finally, the development of LithoGraphX being open, we were able to adapt its rendering engine to manipulate and visualize efficiently high-resolution 2D images (up to 16,384 pixels wide) and add the manipulation tools needed for this protocol.

## 2 Materials

### 2.1 General Remarks

1. This protocol is a modified version of the one presented by Sankar et al.

### 2.2 Equipment

1. Facility to growth plants (i.e., plant growth chamber, phytotron, greenhouse).
2. Bell for vacuum apparatus.
3. Embedding molds (e.g., Molding Trays 6 mm × 12 mm × 5 mm (27304) Plano).
4. Acetate foils.
5. Bi-adhesive tape and super glue.
6. Supporting blocks for cutting (e.g., 1.5 cm × 1.5 cm × 2 cm wooden/plastic blocks).
7. Rotary microtome (e.g., Leica RM2255) with Glass knife holder or Tungsten knife holder.
8. Light microscope with motorized stage and a B/W camera (e.g., Zeiss M2 Imager microscope equipped with a HAMATSU camera C11440) or confocal microscopy with motorized stage (*see* **Note 1**).
9. PC computer running Linux or Windows with a high-end processor such as an Intel Core i7-4770 or i7-4790 and 8 GB of RAM (to estimate how much RAM will be needed, the most expensive algorithm for the processing will need around 30 bytes per pixel). At last, it is highly recommended to have an NVIDIA graphic card with 2 GB of memory, although the protocol should work without.

### 2.3 Solutions and Kits

1. Technovit7100 embedding kit (64709003), Heraeus Kulzer.
2. Fixing solution: 4% Formaldehyde (4971-1, Roth), 1% Glutaraldehyde (G6257, Sigma), 50 mM, 0.1% Triton, in Phosphate buffer pH 7.2.
3. 10%, 30%, 50%, 70%, 96% ethanol.
4. Solution A: 10 mL of Technovit7100 cold-curing resin, 0.1 g Technovit 7100 Hardener I, 200 μL of PEG 400 (P0813, Duchefa).
5. Embedding solution: (16:1 solution A: Technovit 7100 Hardener II (v/v)) add 2 μL Technovit 7100 Hardener II to 32 μL of and mix well (*see* **Note 2**).
6. 0.1% Toluidine blue.

### 2.4 Software

1. LithoGraphX 1.1.6 or greater ([14]; *see* **Note 3**).
2. Fiji (http://fiji.sc/Fiji; [15]).

3. Python (http://www.python.org).
   (a) NumPy/SciPy (http://www.scipy.org).
   (b) Pandas (http://pandas.pydata.org).
   (c) scikit-learn (http://www.scikit-learn.org).
4. If, as recommended, you have an NVidia video card, you need to install the Cuda-enabled drivers for your graphic card (*see* **Note 3**) (http://www.geforce.com/drivers).

## 3 Methods

### 3.1 Experimental Design

This protocol is optimized for Arabidopsis (Columbia strain) hypocotyls growing in soil in a (16 h light) plant growth chamber for 14, 21, 28, and 35 days. In principle it can be adapted to any other growth condition, developmental stage, organ or species. We suggest making a pilot experiment to choose the most relevant developmental stages and test the growth conditions. We recommend growing a sufficient number of plants, as not all samples will be suitable for analyses for technical reasons. For example, starting with 50 plants per time point to randomly, collecting 40 hypocotyls per developmental stage and obtaining 25–30 good cross sections (per time point) should lead to the successful analysis of 20–25 plants per developmental stage.

### 3.2 Embedding in Plastic Block and Sectioning

The organs of interest (e.g., root, hypocotyl, stem) are embedded in plastic blocks using the Technovit7100 kit prior to sectioning with a rotary microtome.

1. *Fixation*: quickly collect samples in water and transfer in fixing solution. The fixation step is performed at room temperature for a minimum of 3 h until 24 h (*see* **Note 4**).
2. *Dehydration*: remove the fixing solution and rinse the samples with water twice. Then, progressively dehydrate the samples with successive incubations: 30 min in 10% ethanol, 30 min in 30% ethanol, 30 min in 50% ethanol, 1 h in 70% ethanol (at this step it is possible to store the samples at 4 °C for 3 weeks) and three times 1 h in 96% ethanol.
3. *Pre-infiltration*: pre-infiltrate the samples overnight in 50% ethanol 50% Solution A.
4. *Infiltration*: infiltrate samples in solution A for a minimum of 1 day and a maximum of 3 days.
5. *Embedding*: for very tiny samples such as young roots/hypocotyls, we recommend a two-step embedding protocol as directly embedding in the final mold may result in samples with random orientation. Handcrafted embedding cassettes with different thickness can be easily made with acetate foils and bi-adhesive

tape (*see* **Note 5**, Fig. 1d). Place the samples in the embedding cassettes/mold and align them according to the preferred direction. Carefully remove solution A and replace it with the embedding solution. Seal the cassette/mold with acetate foil and let it polymerize over-night (*see* **Note 6**). Open the embedding cassettes and remove the plastic layer. Align several plastics layer (2–5) on the top of each other and glue them together with the embedding solution to obtain a pile. Place the pile in the embedding mold in the desired orientation and cover it with the embedding solution. Seal the mold with a piece of acetate foil and let it polymerize overnight (*see* **Note 6**). Plastic blocks can be stored for years in a dry environment.

6. *Sectioning*: Trim and glue the blocks to a support (e.g., wooden/plastic block) for sectioning. Cut them with a glass or tungsten knife (for hard woody samples) at 3–7 μm. Collect the sections in warm water, place them on a glass slide and let them dry on a heating plate at 40 °C. Stain the slides with 0.1 % of toluidine blue for easier visualization of the samples (the staining per se is not necessary for image acquisition, however if the pictures are acquired in bright field mode, a soft staining helps to highlight the cell borders).

### 3.3 Image Acquisition

High-resolution images are a prerequisite for success of this work (*see* **Note 7**). Modern microscopes typically cannot image at high resolution an object as large as the cross section of the hypocotyl as one image. Instead, we rely on tiling (e.g., collecting images organized into a grid that covers the whole object) and stitching (e.g., merging the tiles into a single high resolution image). The tiling requires a microscope with a motorized stage, while the merging can be done either with the software of most microscopes, or with FiJi (please refer to the microscope manufacturer for detailed instructions on tiling/stitching). The number of tiles is selected according to the size of the sample and to the chosen magnification. For example for a circular sample of 0.5 mm of diameter (hypocotyl section of 21-day-old Columbia Arabidopsis plant), 16 tiles (4 × 4; each tile 2048 × 2048 pixels; 63× objective) are sufficient (*see* **Note 7** and Fig. 1).

LithoGraphX is optimized for high-resolution 16-bit grey scale images, thus we highly recommend acquiring 16 bit images with a black and white camera. The required format for image processing is TIFF or OME-TIFF. If the microscope's software does not offer an OME-TIFF export, it is recommended to save the image in the custom format of the microscope supplier (i.e., for Zeiss *.czi* for leica *.lif*) and convert it to TIFF later using FiJi.

### 3.4 Organization of the Data

To simplify the automation of the classification process, it is highly recommended to choose a naming convention for the files

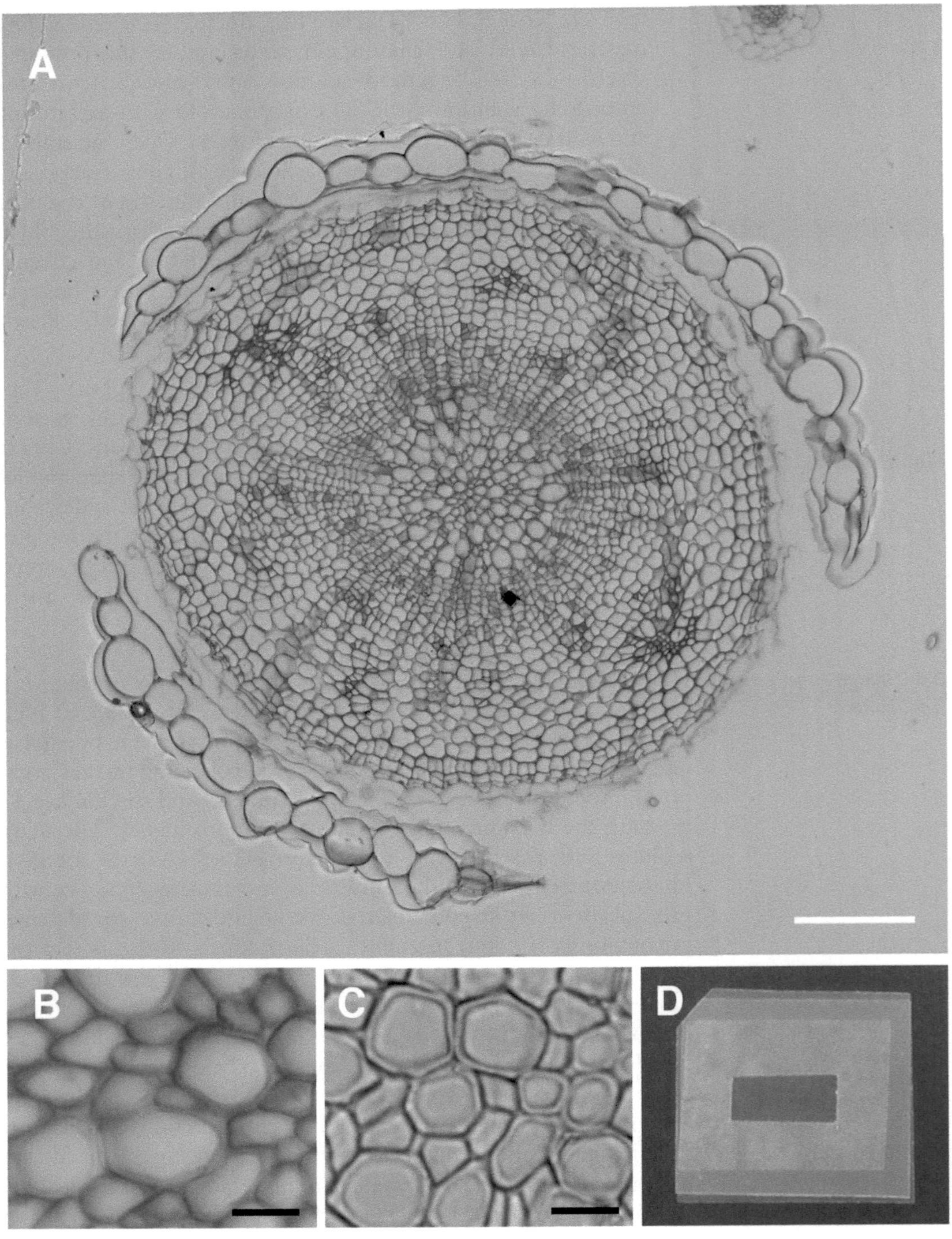


**Fig. 1** Image acquisition. (**a**) Example of a row (stitched) image. Cross section of a 21-day-old Arabidopsis (Columbia) hypocotyl. The image was acquired with a 63× objective in bright field, as a panel of 4 × 4 tiles of 2048 × 2048 pixels and merged with the stitching software of the ZENpro2 software. (**b**) Detail of (**a**) showing xylem vessels. (**c**) Examples of a picture in which xylem vessels will not be segmented correctly with the default parameters. (**d**) Handcrafted cassette for embedding small samples. White bar: 100 μm black bar: 10 μm

generated during the process and to apply it in a systematic manner. We recommend the creation of a separate folder for each stage requiring the creation of a different machine learning classifier. Within this folder, a sub-folder can be created for each acquired image and all files associated with it, allowing all files related to a given sample to be in their own folder. Whether you choose to create sub-folders or not, it is important to choose suffixes for the different types of files you are going to generate. All files related to a given image should share a common prefix and differ only in their suffix. For the following processes, here is the list of suggested suffixes:

- Original image: .tif
- Cell mesh: .ply
- Hand-defined cell types: -training.csv
- Feature file: -features.csv
- Estimated cell types: -celltypes.csv
- Feature files with estimated cell types: -features-typed.csv

### 3.5 LithoGraphX

All image processing and analysis will be performed in the open source software LithoGraphX. For installation, tutorials and basic commands *see* **Note 3**. Designed to work with large stacks or images, LithoGraphX does not have an undo button. And although there is a mechanism to set a single checkpoint (using the 'Main' and 'Work' versions of the stack), it is highly recommended to save the various intermediate steps of the processing, in particular before manual edition phases. For the explanation about the naming of user interface elements in this protocol, *see* **Note 8**.

### 3.6 Image Visualization

1. Start LithoGraphX.
2. Drag and drop the image into the visualization area (which is black by default). The image will be displayed by default in green scale (*see* Fig. 2a and **Note 9**). If some images are already opened, the new image will be loaded in the active store.

### 3.7 Image Processing

The goal of the image processing is to extract the locations and shapes of the cells in the tissue.

1. *Segmentation*. Segmentation is the process in which pixels of the image are associated to objects of interest (in our case cells). The images generated by this protocol have dark cell walls and bright cell interiors. A common method to segment such images is called the Watershed algorithm (for a review on Watershed see [16]).
   (a) Select the **`[Stack]Segmentation/Segment Session`** process, in the bottom-right corner, the parameters rela-

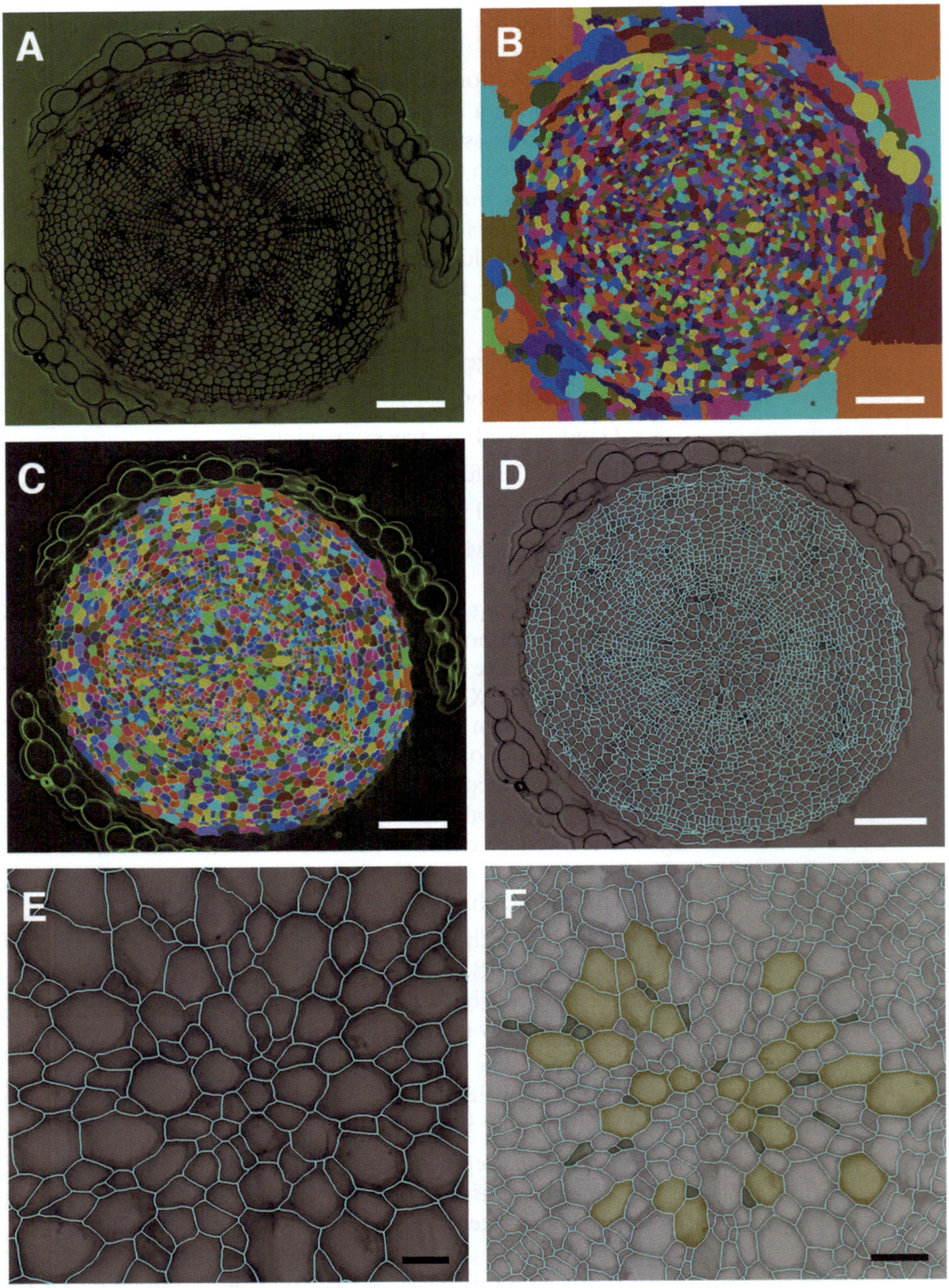


**Fig. 2** Image processing and segmentation in LithoGraphX. All panels display the same image (the same image is presented in Fig. 1a) at different steps of the pipeline. (**a**) Example on how an image is visualized in LithoGraphX. (**b**) Example of a segmented image. (**c**) Example of an image in which labels not corresponding to region of interest have been removed. (**d**) Example of an overlay image: in the background the original picture, in cyan the mesh of the segmentation. (**e**) Detail of (**d**) showing the xylem area. (**f**) Example of cell type labeling with the "parent" function (yellow = xylem vessels, Green: xylem parenchyma). White bar: 100 μm, black bar: 10 μm

tive to this process will be displayed. The default parameters should work for most of the images (*see* **Note 10**), the only option that has to be changed is the last one: **`Invert`**. As the images generated by this protocol have dark cell walls and all processes expect bright ones you need to set it to **`Yes`**.

(b) Click the **`Go`** button to run the process. When the process starts, there will be a warning that the main stack will be modified. If you have made modifications to it before, you may want to save the stack to a file before the process, if not just press **`Yes`**. The segmented image will be displayed in the **`Work`** stack as a color basin image (*see* Fig. 2b). For troubleshooting on the segmentation process refer to **Note 10**.

(c) We recommend saving both the segmented and the preprocessed image at this point (*see* **Note 11**).

2. *Remove non-cells outside the tissue.* The Watershed methods will cover the whole image with cells, including parts outside the sample. There are three ways to delete labels that are not part of the section. It is possible to remove them manually; with the **`[Stack]Segmentation/Remove Labels in Shape`** process; or by subtracting a mask.

(a) To remove them manually, use the ***`Delete picked label in volume`*** tool (scissor pipette icon) and click on the cells to be deleted while pressing the 'Alt' key.

(b) By default the process **`[Stack]Segmentation/Remove Labels In Shape`** will remove any label that is not located in a sphere of size 1 μm and centered on the center of the stack. This process is very fast. So the best way to find which sphere size is needed is to start with a big size and reduce it progressively (It will delete whole labeled regions if their center is not within the defined shape). For example for a Col 21-day-old hypocotyl section, it is good to start with a size of 300 and to reduce progressively by 10 units (*see* Fig. 2c).

(c) To subtract a black and white mask, load the segmented image in the **`Main`** stack and the mask in the **`Work`** stack, then select the **`[Stack]Morphology/Apply Mask to Labels`** process. The default parameters will subtract the white area, if the area to be erased is black in the mask set the **`Invert`** parameters to **`Yes`** (to generate the mask, *see* **Note 12**).

3. *Create a Mesh.* Once the cells have been segmented (i.e., identified pixel by pixel in the image), we want to extract their shape for further processing. In LithoGraphX, a cell mesh defines a set of non-overlapping polygons, each one describing the shape and position of a unique cell.

The mesh file can be uploaded on the original picture, to create an overlay image (*see* Fig. 2d and e). This is the best moment to evaluate segmentation accuracy (in most cases we have a segmentation accuracy of more than 90%).

(a) Select the **[Mesh]Cell Mesh/Cell Mesh From 2D** Image process, setting the **Min Length** parameter to 1 μm.

(b) Back to the **Main** tab, uncheck the **Surface** check box, check the **Mesh** check box and select **Cells** in the **View** drop box. Also, uncheck **Work** and check **Main**. The inverted original image with the contour of the segmented cells in cyan should be visible.

(c) Save the mesh file clicking on the save icon in the **Mesh** group box (Meshes are saved to as .PLY file).

4. *Correct mis-segmented cells.* In LithoGraphX, it is possible to manually correct errors in the segmentation (*see* Fig. 3), to correct over-segmented cells *see* **Note 14**, to correct under-segmented cells *see* **Note 15**.

5. *Mark cell identity.* To manually mark cell identity, we are going to use the notion of "Parent" cell, but instead use it to mark cell type.

(a) In the **Main** tab, uncheck the **Work** check box, check the **Surface** check box and choose "Parents" in the surface drop box. You should now see the exact same cells, but without their color. To better visualize the cells it is possible to reduce the opacity in the **Surface** box.

(b) Use the menu entry **Labels & Selection→Change current label**... In the dialog box, choose a number strictly greater than 0 (0 is the default number for unlabeled objects). This number will be the identifier for the cell type (i.e., we used 3 for xylem vessels, 4 for phloem bundles, 5 for cambium, 6 for periderm, 7 for xylem parenchyma, 8 for phloem parenchyma, and 11 for intercellular spaces).

(c) Select the ***Replace clicked surface label or parent with current one*** tool (2D bucket icon), Alt-click on all the cells that correspond to the given cell type. Cells of the same tissue will become colored with the same color. Repeat the process for all cell types, leaving non-cells unmarked (*see* **Note 16** and Fig. 2f).

(d) The results of the labeling will be integrated in the mesh file upon Saving the Mesh. Alternatively the results can be saved selecting the process **[Mesh]Lineage Tracking/Save Parents**. This generates a CSV file with in the cell label in the first column and the cell type identifier in the second column.

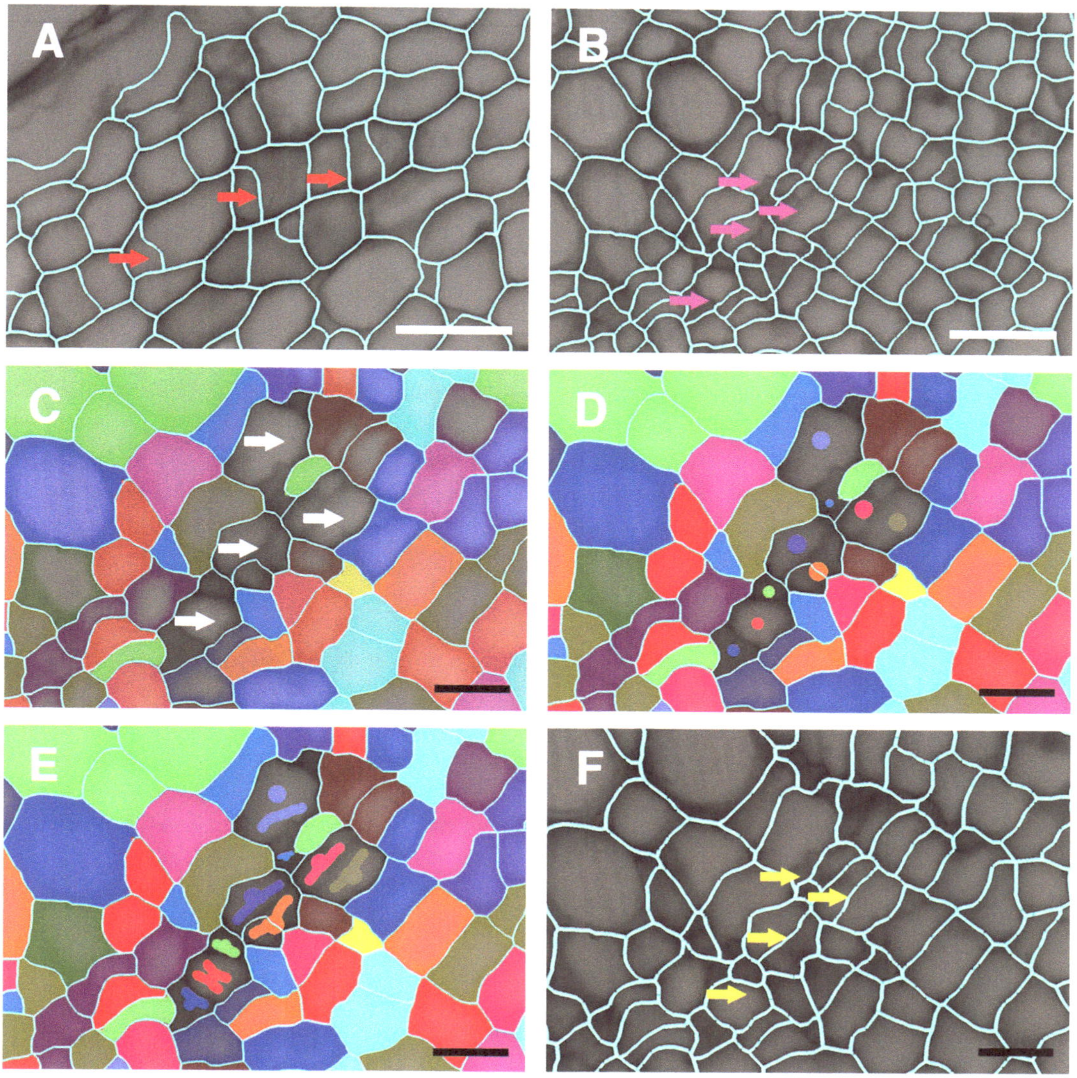


**Fig. 3** Manual correction of the segmentation in LithoGraphX. (**a**) Example of over-segmented cells (*red arrows*). Overlay of the original picture (the same image is presented in Fig. 1a) with in cyan the mesh of the segmentation. (**b**) Example of under-segmented cells (*pink arrows*). Overlay of the original picture (the same image is presented in Fig. 1a) with in cyan the mesh of the segmentation. (**c**–**f**) Manual correction of the under-segmented cells presented in (**b**). (**c**) The labels of the under-segmented cells are erased (*white arrows*). (**d**) New seeds are placed at the center of the cells (*color dots*). (**e**) The borders of the cell are drawn to guide the segmentation. (**f**) New overlay showing that the segmentation has been corrected (*yellow arrows*). White bar: 20 μm and black bar: 10 μm

### 3.8 Computation of Cellular Features

Cellular features (or descriptors) provide geometric characteristic of cell shapes and relative position of cells in the section. These features can be then used to automatically classify unknown cells into the cell types defined earlier. We selected the 16 cellular descriptors used by Sankar et al. and implemented in LithoGraphX (*see* Table 1). It is possible to compute the cellular descriptors

**Table 1**
**Cell descriptors/features**

| Feature | Unit | Dimension | Description |
|---|---|---|---|
| Local position | μm | 3 | Position of the cell in the reference frame of the image |
| Global position | μm | 3 | Position of the cell in the global reference frame. In this protocol, the sample is moved so its center is at position (0,0,0) |
| Polar radius | μm | 1 | Distance from the cell to the center of the sample (e.g., norm of the global position) |
| Polar theta | radian | 1 | Angle between the *x* axis and the line going from the center of the sample to the cell |
| Major axis | μm | 1 | Length of the major axis of the best fit ellipsis. The best fit ellipsis is defined as the ellipsis minimizing the square distance to the cell shape |
| Minor axis | μm | 1 | Length of the minor axis of the best fit ellipsis |
| Eccentricity | | 1 | Eccentricity of the best fit ellipsis |
| Major axis theta | radian | 1 | Orientation of the major axis of the cell with respect to the *x* axis |
| Area | $\mu m^2$ | 1 | Area of the cell |
| Perimeter | μm | 1 | Perimeter of the cell |
| Average radius | μm | 1 | Average radius of the cell (e.g., average distance from the center of mass of the cell to its perimeter) |
| Minimum radius | μm | 1 | Minimum distance from the center of mass of the cell to its perimeter |
| Maximum radius | μm | 1 | Maximum distance from the center of mass of the cell to its perimeter |
| Cell extension | μm | 1 | Geometric mean of the minimum and maximum axes of the best fit ellipsis |
| Incline angle | radian | 1 | Angle between the major axis of the best fit ellipsis and the line from the center of the sample to the center of the cell |

Description of the features used in this study. Features were calculated as in [9]

individually for each mesh, but we also provide a process that automates this task. In this protocol, we describe how to use the automated process.

1. Clear Work and Main stack using the menu entries **`Stack→Stack 1→Reset`** and **`Stack→Stack 2→Reset`**.
2. Launch the **`[Mesh]Cell Mesh/Cell Classifier/ Generate Classifier Features`** process. A configuration dialog box appears when you do. The configuration dia-

log box is split in three zones: the input, the output and a preview of the files to be processed.

3. First, select the folder that contains all the mesh files. If the meshes are in different subfolders, you will need to check the **`Recursive`** check box.

   Enter the pattern that matches the file names of the meshes. The default is probably correct, unless there are meshes you do not want to use (*see* **Note 17**).

4. If cell types have not been saved with the mesh but instead they have been saved in a separate CSV files, check the **`Cell Type Suffix`** check box and specify the suffix used to name those files. This means the files containing the cell types must have the same prefix as the meshes (*see* **Note 17** and **step 5** of Subheading 3.7).
5. For this protocol, the feature process needs to be "Cell_Shapes_Features". It is possible to create and use other process, as long as the first parameter is the name of the file to generate.
6. Specify the suffix for the output CSV files containing the descriptors (i.e., -features).
7. Double check that the list of input files and the names of the generated output files are corrected.
8. Validate the dialog box.
9. A CSV file has been created in the directory of the corresponding mesh.

### 3.9 Computing the Classifier

Once we computed the cellular descriptors and associated the known cell types, we can compute the classifier, which will be later used to automatically recognize cells from their shapes.

1. Launch the **`[Mesh]Cell Mesh/Cell Classifier/Create Cell Classifier`** process. This will open a configuration dialog box.
2. Specify the folder containing the meshes and computed feature files. This should be the same folder you used to compute the features themselves, also checking the **`Recursive`** check box if needed.
3. Specify the cell range you are interested in. This can be useful if you want to exclude some cell types. For example, we decided to label 100 badly segmented cells, so we can test if the classifier can recognize them. If it cannot we can simply ignore them there.
4. Specify the suffix of the feature files. Note that the mesh file pattern and feature file suffix must be the same as in the previous step.
5. By default all machine learning methods are selected. At least one has to be selected, for our images we found that **`Support Vector Machine`** gives the best results. Reserve a percentage

of the data set to compare them (for example 10% or 20%, depending on how many cells you have in total). Default parameters work fine in the majority of the cases for this protocol. It is still possible to optimize the parameters but it requires deep knowledge in machine learning.

6. Specify the output file name for the summary of the machine learning process, which will include predicted precision and best parameters found and the output file name for the classifiers (*see* **Note 18**).
7. Validate the dialog box.
8. The summary and the Classifier file will be in the base directory. The summary file comprises a confusion matrix (for each predicted cell types shows the real cell type), a classification report and a score (see Fig. 4a). The score and the accuracy of the classifier depend on a lot of factors, the size and the accuracy of the training set, the accuracy of the segmentation and on the number of classes (cell types).

### 3.10 Automate Cell Type Recognition Using the Classifiers

Once the classifiers have been computed, they can be used to estimate the cell type of a given image. Classifiers computed for determinate developmental stage may not work for other developmental stages or for images that were acquired in a different way. We suggest computing classifiers for all developmental stages. Filtering post machine learning to improve the accuracy of cell type is beyond the scope of this protocol as it is very specific to the developmental stages, to the organ of interest and to the aim of the study; for the Arabidopsis hypocotyl refer to [9].

1. Launch the **`[Mesh]Cell Mesh/Cell Classifier/Use Cell Classifier`** process. It will open a configuration dialog box.
2. Select the base folder containing all the features files for the meshes you want to identify (the feature files can be computed following Subheading 3.8).

   As before, specify the mesh file pattern and the suffix for the feature files.
3. Specify the name, and possibly the full path, of the file containing the classifier (*see* **Notes 17** and **18**).
4. The process will save a cell type file for each mesh, which you will be able to load later on the image. If the **`Feature suffix`** check box is checked, it will also generate a CSV file for each

**Fig. 4** (continued) of Arabidopsis hypocotyl (21 days) in which cells were automatically recognized using the classifier shown above (yellow = xylem vessels; purple = phloem bundles, cambium = cyan, periderm = brown, xylem parenchyma = dark green, phloem parenchyma = blue, red = intercellular spaces). (**c**) Same image as in (**b**), in which the cell types have been manually corrected after the Use Cell Classifier process. (**d**) Detail of (**c**) showing the xylem area. White bar: 100 μm, black bar: 20 μm

**A**

```
Machine learning process
========================
Selected classifier:  Support Vector Machine


Support Vector Machine
----------------------

Best score with k-fold: 0.93899057127
Parameters:
C : 150.116954624
gamma : 0.00624272274944


Evaluation using test set with 450 elements
"""""""""""""""""""""""""""""""""""""""""""

Confusion matrix:
Predicted    3   4   5   6  7    8  11  Sum
Real
3           55   0   0   0  0    0   0   55
4            0  51   0   0  0    2   6   59
5            0   0  74   0  0    0   0   74
6            0   2   0  72  0    1   0   75
7            0   0   0   0  6    0   0    6
8            0   0   0   1  0  153   0  154
11           0   6   0   1  0    0  20   27
Sum         55  59  74  74  6  156  26  450

Classification report:
             precision    recall  f1-score   support

          3       1.00      1.00      1.00        55
          4       0.86      0.86      0.86        59
          5       1.00      1.00      1.00        74
          6       0.97      0.96      0.97        75
          7       1.00      1.00      1.00         6
          8       0.98      0.99      0.99       154
         11       0.77      0.74      0.75        27

avg / total       0.96      0.96      0.96       450


Score:  95.7777777778 %
```

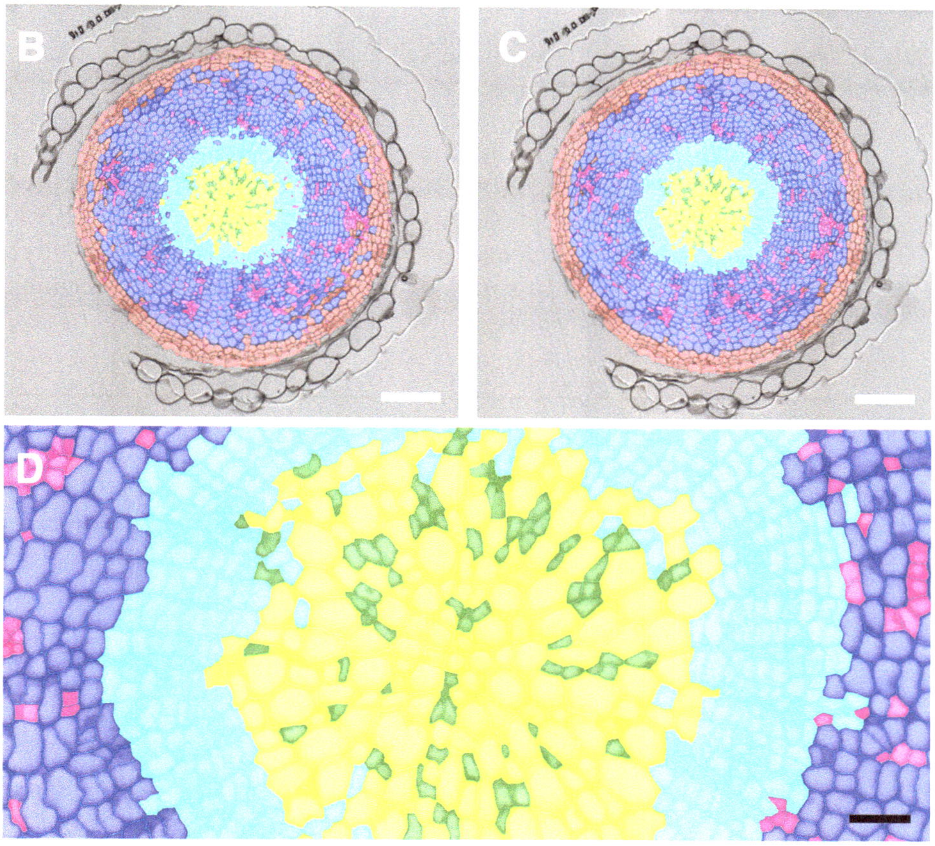


**Fig. 4** Automated cell recognition through machine learning in LithoGraphX. (**a**) Example of a summary file generated with the Create Classifier process (3 = xylem vessels; 4 = phloem bundles, 5 = cambium, 6 = periderm, 7 = xylem parenchyma, 8 = phloem parenchyma, 11 = intercellular spaces). (**b**) Example of a cross section

mesh with the features and cell identity. You need to specify a suffix for those CSV files.

5. It is possible to specify a minimum cell area. This is useful if there are a lot of inter-cellular spaces, which will be identified as very small cells. This parameter allows the algorithm to ignore those spaces, which will then be left unlabeled (unlabeled cells have by default 0 as cell id/parent number).
6. Double check that the list of input files and the names of the generated output files are corrected.
7. Validate the dialog box.
8. To visualize the results, load a mesh, and then use the **[Mesh] Lineage Tracking/Load Parents** process to load the cell type file. The labeling of all cells should then be displayed (see Fig. 4b).
9. It is possible to manually correct misidentified cell simply changing the current label in the **Label & Selection→Change current label...** menu and using the ***Replace clicked surface label or parent with current one*** tool (2D bucket icon) with the parent shown to change the labeling (*see* **Note 16** and Fig. 4c and d).

### 3.11 Data Analyses and Integration

The CSV file with the cell type and the features contain all the information to calculate most important parameters to describe secondary growth progression (*see* **Note 19**). Proliferation rate, total organ area, number of cells per tissue can be easily estimated with Excel (see **Note 20**) or with the statistical program of your choice. For example total area of the sample (i.e., hypocotyl area) can be easily calculated summing the area of all the objects (cells and intercellular spaces), xylem area can be estimated summing the area for all xylem cells.

## 4 Notes

1. If a light-microscope with motorized stage is not accessible, it is possible to use a confocal microscope with motorized stage and acquire images in transmitted light modus with the stitching/tiling function.
2. Embedding solution should be prepared fresh few minutes before use.
3. LithoGraphX can be downloaded at http://www.lithographx.org. The list of dependencies and installation instructions can be found there: http://lithographx.readthedocs.org/en/latest/installation.html.

   There are a few options to install LithoGraphX and its dependencies. We recommend installing the binaries of

LithoGraphX and using Anaconda to install the python dependencies. Basic instruction, tutorials and protocols are also available on the web site.

In the Help menu of LithoGraphX there is a link to the User Manual, a list of Mouse and Keyboard commands, and the description of every process.

The tools present in the left vertical bar are active only if the ALT button is pressed.

4. For very hard samples we recommend to vacuum infiltrate for 1 h and fix them over night at room temperature.
5. Handcrafted embedding cassettes are prepared with acetate foil and bi-adhesive tape. Cut 3 rectangular pieces of 2 cm × 3 cm of acetate foil. Place in the center of an acetate rectangle a piece of bi-adhesive tape of 1 cm × 2 cm and glue on the top of it a second acetate rectangle. Place another piece of bi-adhesive tape, in the center of the glued acetate rectangles (the pieces of tape should be aligned; at this step adding more than one piece of bi-adhesive tape will create a deeper cassette). With a sharp scalpel/blade cut out a rectangle of 0.5 cm × 1 cm (the window should be cut in the center of the tapes), creating a window. Place the last acetate piece. The cassette is now ready to use.
6. Seal completely your embedding cassettes and molds as oxygen prevents the polymerization of the resin.
7. How images are acquired deeply influences the success of the segmentation, here we give some tips, and some examples, but which method (i.e., Bright field versus DIC) may work better depends on the microscope, and on the samples. The center of the sample should coincide with the center of the image. Cell borders should be clearly visible and complete, (any gap in the cell border could result in fused cells during the segmentation). Maximize the color difference of the cell border versus the inside of the cell (i.e., cell borders should be uniform black, and the inside of the cell grey). A soft blue of toluidine staining will help to see the cell border without altering the color inside the cells. A too strong staining may result in shadows and non-homogeneous color inside the cell (*see* Fig. 1). It may happen that during the stitching process some shades of the different tiles are visible in the merged picture. If the shadows are not too strong and do not create any black line, they should not interfere with the segmentation process. In case the shadow is really pronounced it is possible to remove it, subtracting a background picture (a picture taken with the same setting but without sample). For example, the images used in this study were acquired with a Zeiss M2 Imager microscope (with the ZEN Pro 2 blue edition software), in bright field with a 63× objective. Stitching was performed with the default stitching

parameters (no stitching while acquiring; 10% overlap). Merging of the tiles was performed after acquisition with the image-processing tool of the ZEN Pro 2 software and an automatic shadow correction is simultaneously applied (see Fig. 1a). Successively, images were exported using the method OME. TIFF export (this format is directly supported in LithoGraphX).

8. LithoGraphX provides command through four different graphical elements: an application menu, toolbars, a list of processes and a series of buttons and checkboxes.

   - In this protocol, we use distinct conventions to distinguish between these: Menu entries will be referred as **Menu → Submenu → Entry**.
   - Processes will be referred as **[Type]path/to/process name**. Where the type indicates which tabs the process will be found in, the path indicates the location in the hierarchy of processes and the process name is the unique identifier of the process.
   - Tools in a tool bar will be referred with the name of the tooltip like this: ***Delete picked label in volume***.
   - Other GUI (Graphical User Interface) elements will be located explicitly, and the name appearing on the interface will be highlighted. For example, we refer to the **Main** group box in the **Stack 1** tab.

9. It is possible to visualize the image in grey scale and to adjust the brightness with the transfer function.

   (a) Click on the rainbow icon, in the **Main** tab, to open the **Transfer Function editor**. A histogram with intensity values of the images becomes visible.

   (b) Adjust the brightness, moving the black arrowhead in the histogram.

   (c) In the drop box menu of **Predefined transfer functions** select **Scale gray** and then click on **OK**.

10. In case of failure of the segmentation or in case of numerous segmentation mistakes, we suggest: (a) changing the default parameter of the Gaussian Blur (**BlurX, BlurY**) radius. If large cells are under-segmented, it could be because of "holes" in their walls. In this case, increasing the blur radius can help. If small cells disappear, it may be because the blur is too large and they are replaced with a thick cell wall. In this case, the blur radius should be decreased. In both cases, the radius should not be much larger than two or three times the thickness of the cell wall in μm. (b) A common cause of over-segmentation is the amount of background noise inside cells. To avoid this problem you can increase the level parameter of the Watershed process

(**WatershedLevel**). This parameter corresponds to the darkness of the cells' background and ranges from 0 to 65,535. If too low, cells will be over-segmented. If too high, cells will be under-segmented. (c) Subtracting a background picture if the mistakes are due to the shadows present in pictures caused by the tiling. (d) Applying a filter to remove noise if the inside of the cells contains shades, or it is not homogeneous. Note that the process requires 42 bytes of memory per pixel in the image, so for an image of 8000 × 8000 pixels (size of the Col 21-day-old example) you will need 2.6 GB.

(a) Select **[Stack]Morphology/Opening**, run the process with 1 pixel on each dimension.

(b) Select **[Stack]Morphology/Closing**, run the process, with the same parameters (the order of the processes should not matter).

(c) Select **[Stack]Morphology/Sieve Filter**. The first parameter should be **Median**, the second parameter should be an area in $\mu m^2$, a bit smaller than the smallest cell you want to segment (we used 3 $\mu m^2$ in our examples) and the last parameter should be **No**. The progress bar will display how many extrema there are. If there are too many (>1,000,000), then you should cancel and run again the opening and closing with a larger radius (2 or 3), otherwise it will be really slow.

11. A shortcut to save images/meshes in LithoGraphX, is to click on the save icon in the correspondent group box (Main, Work, Surface). For example to save the pre-processed image located in the Main store, go to the **Main** tab, and then the **Stack 1** tab, from there, click on the save icon in the **Main** group box. To save the segmented image click on the save icon in the **Work** group box, just below the **Main** group box. Alternatively it is also possible to save images/meshes using the corresponding process: **[Stack]System/Save** or **[Mesh]System/Save** or the menu entries such as **Stack→Stack 1→Main→Save** and **Mesh→Mesh 1→Save**.

12. A black and white mask defining the area of interest of the picture can be draw in FiJi or any imaging software, saved as a TIFF file and loaded in LithoGraphX. It is also possible to directly draw the mask in LithoGraphX: in the **View** tab, in the **Stack editing** section check the **Fill** box, then in the **Main** tab select the **Work** stack, select the ***Set cylinder of voxels under the cursor to the current label or pixel edit value*** tool (the work stack should be clear prior drawing) and draw ALT-clicking on the image. To fill the mask, select the ***Replaced clicked volume label with current one*** (3D bucket) tool and ALT-SHIFT-click on it (*see* **Note 13**).

13. Pressing the SHIFT key convert the ***Replaced clicked volume label with current one*** tool in the ***Flood fill from clicked point*** tool.
14. Over-segmented cells (i.e., one cell is divided in two labels, see Fig. 1) (*see* Fig. 3a) can be easily fixed. Select the ***Pick a label from the volume*** (3D pipette icon) tool, Alt-click on one label of over-segmented cell (the color of the picked label will appear on the top upper corner, and the number on the label will be written at the bottom left corner), then select the ***Replace clicked volume label with current one*** tool (3D Bucket icon) and ALT click on the other labels belonging to that cell. Re-create a mesh as described in **step 3** of Subheading 3.7, now the cell should have only one label.
15. Correcting under-segmented cells (more than one cell are identified by one label) (*see* Fig. 3b) is a slightly longer process.
    (a) Erase the label of the under-segmented cells, with the ***Delete picked label in volume*** tool (scissor pipette icon) (*see* Fig. 3c).
    (b) Add seeds in each of the cells to segment: in the **View** tab in the **Stack editing** section, check the **Fill** and the **New Seed** box, back to **Main** view select the ***Set cylinder of voxels under the cursor to the current label or pixel edit value*** tool (3D spray icon), ALT-click to add a seed at the center of each of the cells that are missing, small disks of different color will appear (*see* Fig. 3d).
    (c) After seeding all the under-segmented cells, fill the non-labeled space around the tissue to prevent outer cells from bleeding. This can be achieved placing another seed in the non-labeled part and filling the rest of the non-labeled space by ALT-SHIFT-click with the ***Replace clicked volume label with current one*** tool (3D bucket) (*see* **Note 13**). To guide the segmentation of these cells, it is possible to delimit the cell borders: with the ***Pick a label from the volume*** tool (3D pipette), select one seed, in the **View** tab in the **Stack editing** section, check the Fill and uncheck the **New Seed** box, back to **Main** view select the ***Set cylinder of voxels under the cursor to the current label or pixel edit value*** tool (3D spray icon), ALT-click to draw the border of the cell (*see* Fig. 3e).
    (d) Select **[Stack]ITK/Segmentation/ITK Watershed** process with default parameters. Before launching the process, be sure that the **Main** check box is checked, and that in the **Main** stack is loaded the inverted image obtained after segmentation (the pre-processed image generated by the process

**[Stack]/Segmentation/Segment Section**). Manually delete the label of the filled outside space around tissue (background) with the ***Delete picked label in volume*** tool (scissor pipette icon). Re-create a mesh as described in **step 3** of Subheading 3.7, now every cell should have a label (*see* Fig. 3f).

16. It there are a lot of intercellular spaces in the sample it is possible to mark them. In doing so, the intercellular spaces will be part of the classification process and the system will try to identify them automatically.

    If you made a mistake in assigning cell identity with the parent function you can easily correct it. In the **Main** tab, uncheck the **Work** check box, check the **Surface** check box and choose "Parents" in the surface drop box. Select the menu entry **Labels & Selection→Change current label...**. In the dialog box, choose 0, if you do not want to assign any identity to that cell or the number corresponding to correct cell type. Select the ***Replace clicked surface label or parent with current one*** tool (2D bucket icon), Alt-click on cell you want to delete the parent or correct it.

17. In all processes related to the classifier, the mesh file is specified as a pattern, which includes the definition of the base name. The base name is the part of the pattern enclosed in curly brackets. This allows the careful construction of the other file names by appending a suffix to the base name. For example, let's suppose the mesh files are called "experiment_date-training.ply", with each file having a unique value for "experiment" and "date", and the feature files should be called "experiment_date-features.csv". Then, the pattern for the mesh should be "{*}-training.ply", as it will select all the files ending with "-training.ply" but only select the part before the suffix as base name, and the suffix for the features should be "-features.csv". On the other hand, if the features should be called "experiment_data-training-features.csv", then the mesh pattern should be "{*-training}.ply", which now captures the whole file name without the extension for the base name, and the feature suffix should stay the same.

18. When saving the classifier file, it is important to remember that its format may change depending on which version of Python or of the scikit-learn library you are using. If you obtain a classifier file from another person, or if you update your system, the process **[Mesh]Cell Mesh/Cell Classifier/Use Cell Classifier** may fail either while loading or using the classifier. Either way, the error will happen before anything is written. If this happens, you will need to regenerate the classifier from the original data.

19. If misidentified cells have been corrected after the **[Mesh]Cell Mesh/Cell Classifier/Use Cell Classifier** process, it

is necessary to add the new labeling to the cell features. You can use the process **[Mesh]/Lineage Tracking/Add Lineage to Table**. Before launching the process, choose a column name to store the new cell types in. When launching the process, it will show a dialog box letting you choose the file to edit. By default, the file will be overwritten with the new column. You can also specify in the parameters the name of the file to write.

20. If you try loading your data using Excel, you will notice that there will be issues: at best strange characters will appear before Greek letters; at worst it will not find the columns. To read a CSV file with Excel, it is recommended to use the "From text" option in the "Data" menu. When opening the file, Excel will then ask a few questions: be sure to select the coma as separating character. If the numbers are not properly recognized (which may happen with some versions of Excel, you will need to open the "Advanced" dialog box of the text import Wizard and select the period as decimal separator rather than the comma. You can check your options with the previous of the table in the import wizard. For more details, check the Microsoft documentation website: on https://support.office.com search "Excel Import TXT or CSV files".

## Acknowledgement

This work was supported by a DFG grant (RA-2590/1-1) and SystemsX.ch, the Swiss Initiative in Systems Biology. We thank Christian Hardtke and Sarah Robinson for critical reading.

## References

1. Ragni L, Hardtke CS (2014) Small but thick enough—the Arabidopsis hypocotyl as a model to study secondary growth. Physiol Plant 151(2):164–171. doi:10.1111/ppl.12118
2. Spicer R, Groover A (2010) Evolution of development of vascular cambia and secondary growth. New Phytol 186(3):577–592. doi:10.1111/j.1469-8137.2010.03236.x
3. Du J, Miura E, Robischon M, Martinez C, Groover A (2011) The Populus Class III HD ZIP transcription factor POPCORONA affects cell differentiation during secondary growth of woody stems. PLoS One 6(2):e17458. doi:10.1371/journal.pone.0017458
4. Etchells JP, Mishra LS, Kumar M, Campbell L, Turner SR (2015) Wood formation in trees is increased by manipulating PXY-regulated cell division. Curr Biol 25(8):1050–1055. doi:10.1016/j.cub.2015.02.023
5. Etchells JP, Smit ME, Gaudinier A, Williams CJ, Brady SM (2015) A brief history of the TDIF-PXY signalling module: balancing meristem identity and differentiation during vascular development. New Phytol 209:474–484. doi:10.1111/nph.13642
6. Kubo M, Udagawa M, Nishikubo N, Horiguchi G, Yamaguchi M, Ito J, Mimura T, Fukuda H, Demura T (2005) Transcription switches for protoxylem and metaxylem vessel formation. Genes Dev 19(16):1855–1860. doi:10.1101/gad.1331305
7. Robischon M, Du J, Miura E, Groover A (2011) The Populus Class III HD ZIP, popREVOLUTA, influences cambium initiation and patterning of woody stems. Plant Physiol 155(3):1214–1225. doi:10.1104/pp.110.167007
8. Meyerowitz EM (1989) Arabidopsis, a useful weed. Cell 56(2):263–269

9. Sankar M, Nieminen K, Ragni L, Xenarios I, Hardtke CS (2014) Automated quantitative histology reveals vascular morphodynamics during Arabidopsis hypocotyl secondary growth. eLife 3:e01567. doi:10.7554/eLife.01567
10. Lehmann F, Hardtke CS (2015) Secondary growth of the Arabidopsis hypocotyl-vascular development in dimensions. Curr Opin Plant Biol 29:9–15. doi:10.1016/j.pbi.2015.10.011
11. Ragni L, Nieminen K, Pacheco-Villalobos D, Sibout R, Schwechheimer C, Hardtke CS (2011) Mobile gibberellin directly stimulates Arabidopsis hypocotyl xylem expansion. Plant Cell 23(4):1322–1336. doi:10.1105/tpc.111.084020
12. Sibout R, Plantegenet S, Hardtke CS (2008) Flowering as a condition for xylem expansion in Arabidopsis hypocotyl and root. Curr Biol 18(6):458–463. doi:10.1016/j.cub.2008.02.070
13. Hall HC, Fakhrzadeh A, Luengo Hendriks CL, Fischer U (2016) Precision automation of cell type classification and sub-cellular fluorescence quantification from laser scanning confocal images. Front Plant Sci 7:119. doi:10.3389/fpls.2016.00119
14. Barbier de Reuille P, Routier-Kierzkowska A-L, Kierzkowski D, Bassel GW, Schüpbach T, Tauriello G, Bajpai N, Strauss S, Weber A, Kiss A, Burian A, Hofhuis H, Sapala A, Lipowczan M, Heimlicher MB, Robinson S, Bayer EM, Basler K, Koumoutsakos P, Roeder AH, Aegerter-Wilmsen T, Nakayama N, Tsiantis M, Hay A, Kwiatkowska D, Xenarios I, Kuhlemeier C, Smith RS (2015) MorphoGraphX: a platform for quantifying morphogenesis in 4D. eLife 4:05864. doi:10.7554/eLife.05864
15. Schneider CA, Rasband WS, Eliceiri KW (2012) NIH image to ImageJ: 25 years of image analysis. Nat Methods 9(7):671–675
16. Vincent L, Soille P (1991) Watersheds in digital spaces: an efficient algorithm based on immersion simulations. IEEE Trans Pattern Anal Mach Intell 13:583–598

# Chapter 11

# Xylem Characterization Using Improved Pseudo-Schiff Propidium Iodide Staining of Whole Mount Samples and Confocal Laser-Scanning Microscopy

**Mario Coiro and Elisabeth Truernit**

## Abstract

An improved pseudo-Schiff propidium iodide staining technique well suited for, but not limited to, the visualization of xylem cell walls in whole mount samples is presented. The pseudo-Schiff reaction results in covalent binding of the fluorescent dye propidium iodide to cell walls. This stable linkage permits the use of clearing agents after staining, which is itself improved following pretreatment of the plant tissue. A subsequent acid alcohol washing step eliminates unbound propidium iodide to reduce background fluorescence. The method can be used for characterizing xylem cell structure in different organs and species without the need for tissue sectioning.

**Key words** Xylem, Propidium iodide, Pseudo-Schiff, Confocal microscopy, Plant tissue clearing, Whole mount

## 1 Introduction

Xylem cells comprise part of the vascular tissue, which is usually deeply embedded in plant organs. Owing to lignification in their cell walls and their characteristic structure, water conducting xylem cells can be easily distinguished from other cell types within the vasculature. The *Arabidopsis thaliana* root has become a model system for studying xylem cell development [1], certainly also because of its thin and transparent root morphology. Xylem cells in Arabidopsis roots can be routinely stained with fluorescent dyes (for example neutral red [2], basic fuchsin [3]), and imaging performed using confocal laser-scanning microscopy (CLSM). However, in other tissue types the position of the vasculature in the center of plant organs, surrounded by thicker and less transparent tissue, can present a challenge in whole mount samples.

A modified pseudo-Schiff propidium iodide (mPS-PI) staining technique can also be used for the visualization of xylem cells [4].

Miguel de Lucas and J. Peter Etchells (eds.), *Xylem: Methods and Protocols*, Methods in Molecular Biology, vol. 1544,
DOI 10.1007/978-1-4939-6722-3_11, 

The fluorescent dye propidium iodide binds covalently to cell walls, thus permitting the use of clearing agents and resulting in a reduction in light scattering from overlying tissue. A hot ethanol treatment [4] or treatment with a sodium hydroxide/sodium dodecyl sulfate (NaOH/SDS) mixture [5, 6] is required prior to staining to improve dye penetration in thicker organs, especially those covered with a waxy cuticle. The latter treatment is especially efficient in clearing and dehydrates the tissue less.

The method presented here is a modified version of the original mPS-PI technique [4], incorporating NaOH/SDS treatment prior to staining, as well as an acid alcohol washing step to remove ionically bound propidium iodide. Collectively, these modifications result in improved penetration of dye, tissue preservation, and a reduction of unspecific fluorescence. This technique has been successfully tested on a variety of whole mount plant tissue samples including ferns, gymnosperms, and angiosperms (*see* Fig. 1 for examples from *Arabidopsis thaliana* and *Fagus sylvatica*).

## 2 Materials

All Solutions Are Prepared with Deionized Water.

1. Clearing solution: 1% (w/v) SDS, 200 mM NaOH.
2. 1% (w/v) periodic acid.
3. Schiff reagent: 1.9 g sodium metabisulfite ($Na_2S_2O_5$), 97 ml water, 3 ml of 5 N HCl (prepare fresh, i.e., up to 1 day in advance).
4. 1 mg/ml propidium iodide solution.
5. 30%, 50%, and 70% (v/v) ethanol.
6. Pseudo-Schiff washing solution: 1% (v/v) 5 N HCl in 70% ethanol.
7. Chloral hydrate solution: 40 g chloral hydrate, 10 ml glycerol, 20 ml water.
8. Hoyer's solution: 30 g gum arabic, 200 g chloral hydrate, 20 g glycerol, 50 ml water. (Stir at 30 °C until gum arabic is dissolved (this can take more than 1 day). Once dissolved, allow the solution to stand for approximately 1 week. Any nondissolved particles will then sink to the bottom of the bottle. Do not shake bottle prior to use!)

## 3 Methods

1. Incubate tissue in clearing solution at room temperature. If fixing is required, rehydrate the tissue prior to clearing (*see* **Note 1**). Thicker tissue (>200 μm) should be dissected

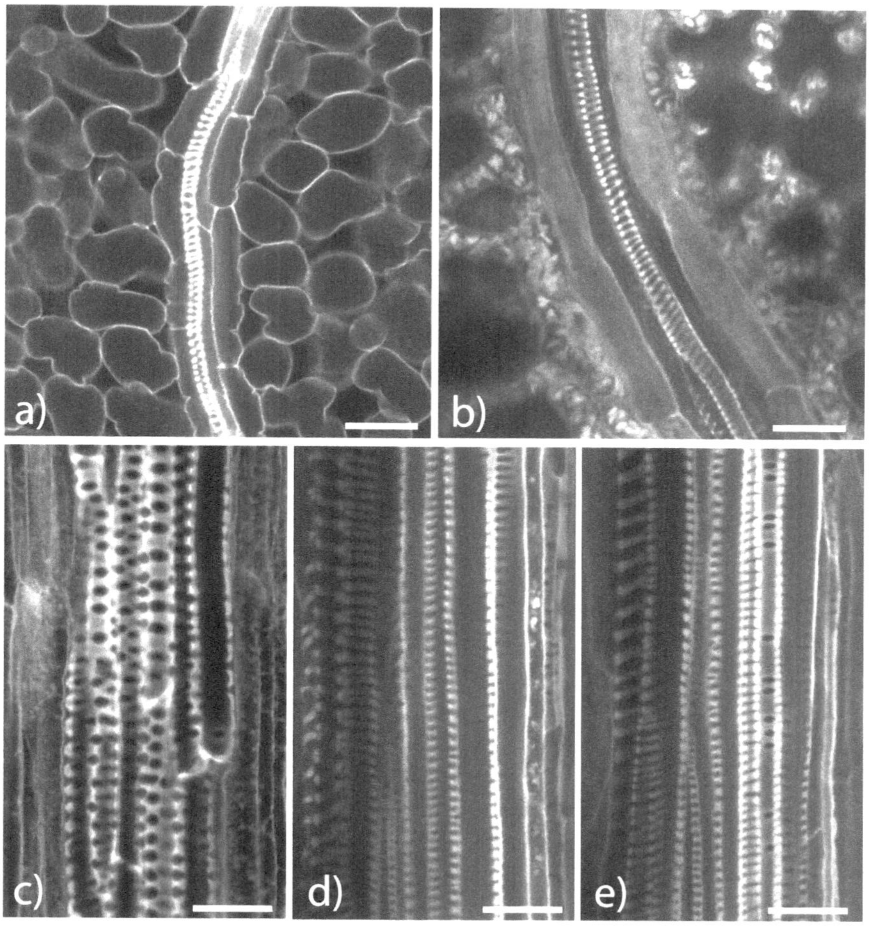


**Fig. 1** Xylem cells of plant tissue treated as described above and imaged using a confocal laser-scanning microscope. All images are optical sections of whole-mount samples. (**a**) Vascular bundle in a cauline leaf of 3-week-old *Arabidopsis thaliana*; (**b**) vascular bundle in the lamina of a *Fagus sylvatica* leaf; (**c**) xylem cells in the hypocotyl of 3-week-old *Arabidopsis thaliana*; (**d**, **e**) xylem cells in the petiole of a *Fagus sylvatica* leaf. Shown are two different focal planes of the same sample. Scale bars: 20 μm

(*see* **Note 2**). Use a vacuum chamber to aid penetration of the clearing solution—floating tissue should sink if successfully treated. Maintain sample at room temperature until the tissue is translucent (*see* **Note 3**).

2. Discard clearing solution and rinse the tissue with distilled water. Repeat this wash four more times.
3. Add 1 % periodic acid and incubate at room temperature for 40–60 min (*see* **Note 4**).
4. Remove periodic acid and rinse five times with distilled water.
5. Add freshly prepared Schiff reagent followed by propidium iodide solution for a final 1:10 ratio. Incubate at room temperature for 1–2 h and for a maximum of 1 week (*see* **Note 5**).

6. Discard the Schiff reagent and rinse sample five times with distilled water.
7. Transfer the tissue to 70 % ethanol through an ethanol series:

   30 % ethanol, 30 min.

   50 % ethanol, 30 min.

   70 % ethanol, 30 min.
8. Add the pseudo-Schiff washing solution. Incubate at room temperature for 10–20 min. The tissue should turn from pink to brown in color.
9. Rehydrate the tissue via an ethanol series (*see* **Note 6**):

   70 % ethanol, 30 min (*see* **Note 7**).

   50 % ethanol, 30 min.

   30 % ethanol, 30 min.

   Distilled water, 30 min.

   Distilled water.
10. Transfer the tissue to chloral hydrate solution. Incubate overnight at room temperature (*see* **Notes 8** and **9**).
11. Mount the tissue in Hoyer's or medium of choice (*see* **Note 9**). For tissues mounted in Hoyer's, allow the medium to solidify under the cover slip for approximately 3 days prior to observation (*see* **Note 10**).
12. The tissue can be observed using a 488 or 514 nm laser.

## 4 Notes

1. Formaldehyde precipitates propidium iodide. Therefore, use other reagent if fixing is required.
2. Please note that the scanning depth of CLSM is usually limited to 200 μm. Moreover, dissection also aids stain penetration (*see* also **Note 5**).
3. The duration of this step is influenced by the organ studied, but all tissue should be visibly translucent before proceeding. For example, 10 min is usually sufficient for small and delicate plant organs or seedlings (for example *Arabidopsis*), whereas overnight or longer treatment is recommended for larger or more coriaceous specimens. However, note that incubation up to 1 week could result in SDS precipitation on the tissue surface. Incubation at 50 °C will significantly speed up the clearing process and may be needed for particularly coriaceous tissues.
4. Tissue oxidation will occur at this step and thus translucency is likely to be lost. It is important to adapt incubation time based on your observations of color change. In our experience, 1 h was sufficient for all our experimental tissues.

5. The incubation time in Schiff/propidium iodide solution is a limiting step in this procedure. Propidium iodide tends to penetrate tissues very slowly, especially when a thick cuticle is present. Different incubation times should be tested and selected in relation to dimension, level of cutinization, or sclerotization. Vacuum infiltration may also help stain penetration in thicker tissues. In bleached or non-chlorophyllous tissues, this is less of a problem and the tissue should turn pink.
6. For particularly robust samples, this dehydration step can be omitted, and the tissue transferred directly from 70% ethanol to the chloral hydrate solution. However, some tissue shrinkage is to be expected.
7. The procedure can be stopped at this step and the tissue stored in 70% ethanol at 4 °C for several months.
8. Chloral hydrate treatment of fragile tissues should be performed directly on the microscope slide. For more recalcitrant tissues the treatment can be undertaken in staining wells. Hardy tissues and materials can be more easily cleared at 50 °C. If samples are on slides, ensure they do not dry out by, for example, storing them in a closed Petri dish. Tissue can be stored for several months in the chloral hydrate solution at room temperature.
9. Different mounting media can be used (for example 100% glycerol, clearing solution (see Subheading 2), glycerine jelly, Histoclear) and specific manufacturer's guidelines should be followed. If cleared with chloral hydrate on slides, carefully remove as much of the solution as possible with filter paper or pipette.
10. Samples can be stored in the dark for several months, but consider that the weight of the cover slip may distort soft tissue samples.

## Acknowledgment

M.C. was supported by an ETH Zurich Research Grant.
We thank Simon Bull for critical reading of the manuscript.

### References

1. Nieminen KM (2004) A weed for wood? Arabidopsis as a genetic model for xylem development. Plant Physiol 135(2):653–659. doi:10.1104/pp.104.040212
2. Dubrovsky JG (2006) Neutral red as a probe for confocal laser scanning microscopy studies of plant roots. Ann Bot 97(6):1127–1138. doi:10.1093/aob/mcl045
3. Mähönen AP, Bonke M, Kauppinen L, Riikonen M, Benfey PN, Helariutta Y (2000) A novel two-component hybrid molecule regulates vascular morphogenesis of the Arabidopsis root. Genes Dev 14(23):2938–2943
4. Truernit E, Bauby H, Dubreucq B, Grandjean O, Runions J, Barthelemy J, Palauqui JC (2008) High-resolution whole-mount imaging

of three-dimensional tissue organization and gene expression enables the study of Phloem development and structure in Arabidopsis. Plant Cell 20(6):1494–1503

5. Wuyts N, Palauqui J-C, Conejero G, Verdeil J-L, Granier C, Massonnet C (2010) High-contrast three-dimensional imaging of the Arabidopsis leaf enables the analysis of cell dimensions in the epidermis and mesophyll. Plant Methods 6:17. doi:10.1186/1746-4811-6-17
6. Nguyen STT, McCurdy DW (2015) High-resolution confocal imaging of wall ingrowth deposition in plant transfer cells: semi-quantitative analysis of phloem parenchyma transfer cell development in leaf minor veins of Arabidopsis. BMC Plant Biol 15(1):431. doi:10.1186/s12870-015-0483-8

# Chapter 12

# Chemical Imaging of Xylem by Raman Microspectroscopy

**András Gorzsás**


## Abstract

Raman microspectroscopic techniques provide highly detailed chemical information about xylem tissue at submicron spatial resolution. The techniques are generally sensitive and they provide a powerful, yet inexpensive way to probe the chemical composition of individual cells or cell wall layers in situ, non-destructively, in a confocal manner, simultaneously detecting all chemical compounds without the need of external agents (label, dyes, stains). Problems with limited specificity in complex chemical matrices such as cell walls may arise, compounded by fluorescence problems. However, these can often be circumvented. In this chapter, the basics of the technique, including a common instrumental setup, together with the general strengths and limitations of Raman microspectroscopy are discussed. Detailed protocols are provided for single point measurements, as well as for fully customizable raster scan mapping, including sample preparation, setup, and measurement steps. The major steps of the data analysis procedure are discussed as well.


**Key words** Raman spectroscopy, Microspectroscopy, Plant, Methods, Chemical composition, Multivariate analysis

## 1 Introduction

Raman spectroscopy is based on the inelastic scattering of light as a result of interaction with molecular vibrations possessing a change in their polarization potential [1]. Therefore, most chemical compounds contain (a set of) Raman active vibrations (*see* **Note 1**). The exact number and energy (given as wavenumbers, in units of $cm^{-1}$) of these vibrations depend on the functional groups the molecule contains and the overall molecular structure and symmetry, including intramolecular and intermolecular features [2]. As a result, every Raman active molecule has a unique spectral fingerprint (the spectrum being the plot of peak intensities as a function of frequency shift from the parent laser line, *see* **Note 2**). In case of chemically pure compounds, these fingerprints can be matched to references for precise identification. In mixtures, however, identification can be limited by severely overlapping or masked features that may result in a lack of diagnostic bands (i.e., bands that would be unique to a certain compound). A further complication is the

Miguel de Lucas and J. Peter Etchells (eds.), *Xylem: Methods and Protocols*, Methods in Molecular Biology, vol. 1544,
DOI 10.1007/978-1-4939-6722-3_12, © Springer Science+Business Media LLC 2017

potential problem with autofluorescence (especially for colored samples/compounds) coupled to low signal strengths. This can hamper the analysis by imposing detection limits and making quantification difficult. While problems attributed to low signal intensities can potentially be overcome by using high intensity lasers and augmented sample preparation (e.g., adding colloidal transition metals to enhance resonance effects [2]), these are not always feasible, depending on the instrumental setup and the nature of the sample. While fluorescence problems can be particularly laborious to solve, Raman spectroscopy represents a fast, inexpensive, sensitive, non-destructive and highly customizable tool for analyzing the entire chemical profile of samples, without the need of external agents (markers, labels, dyes, or stains). In addition, microscopic accessories provide the possibility to map the chemical landscape of samples in situ, with minimal to no extra sample preparation. Since the laser light required for generating the Raman signal can be focused with high precision, these microspectroscopy setups can be operated at high (submicrometer) lateral resolution and in confocal mode, allowing for depth profiling and surface sensitive measurements. By taking advantage of polarizers, even direction sensitive studies can be performed, effectively probing the orientation of all organized (macro)molecules simultaneously, together with the complete chemical profile of the sample.

The current protocol provides detailed steps for single spot measurements (well suited for, for example, high-throughput screens aiming to classify a large number of samples according to their overall chemical profile, or to identify samples with modified chemical composition, i.e. mutant screens) as well as raster scanning an area to produce high resolution chemical maps of tissues, including confocal and polarized measurements. However, depth profiling and high speed/large area coverage methods (using special array detectors, such as Renishaw's Streamline™ setup) are excluded, since they are either less relevant for xylem analysis (depth profiling) or require specific instrumentation (array detectors) beyond the standard Raman microscope. Similarly, specific, augmented techniques (surface enhanced Raman spectroscopy (SERS), atomic force microscopy (AFM) coupled to Raman microspectroscopy, etc.) are outside the scope of the present chapter. Instead, specific sections are dedicated to sample preparation (a key step in acquiring good quality Raman spectra) and basic data analysis steps, to ensure the highest amount of information is obtained from the recorded data. Detailed protocols for hyperspectral data analysis, however, are not included.

### 1.1 Raman Microspectroscopy in Plant Sciences

Vibrational spectroscopic imaging (including FTIR and Raman microspectroscopies) is rapidly gaining recognition in biological and medicinal sciences [3], since it has numerous key advantages over traditional chemical analytical and imaging techniques.

The first key advantage is (time and cost) efficiency. A generic spectrometer with low maintenance can be purchased at relatively low cost, including microscopy accessories and vibration-proof tables. For Raman microspectroscopy, in particular, no additional equipment is required: standard tools available at every plant science laboratory for sample preparation and handling are sufficient in most of the cases (*see* Subheading 2). The only exceptions could be the Raman transparent $CaF_2$ microscopy slides, which are not required for any other (confocal) technique. These, however, are optional and reusable, and therefore do not impose a major cost. Therefore, expanding a plant research laboratory to include Raman microspectroscopic analysis is generally not a major expense. Similarly, running costs are minimal especially when calculated per sample or per hour, and are mostly determined by the working hours (lifetime) of the lasers. Considering that one spectrum can be recorded within a few seconds in most cases, and that every spectrum provides information about the entire chemical composition of the sample, not only about selected target compounds, it is easy to see the time and cost effectiveness of the method. Moreover, samples may not need to be specifically prepared for Raman measurements (e.g., can be done in connection with other microscopy investigations) and the technique is non-destructive, i.e., the measured samples are left unaltered and can be passed on to further processing by other methods.

Drawbacks, however, exist, which may limit the use of Raman microspectroscopy in plant sciences. The major problems include fluorescence, low specificity and difficult quantification. Firstly, plant tissue (including but not limited to xylem) contains material that tend to fluoresce, which can make Raman measurements particularly difficult. There are ways to work around this problem by for example using high laser powers or long exposure times (quenching), highly confocal settings, or switching to ultraviolet/near infrared lasers instead of visible ones. However, none of them are guaranteed to succeed on all samples (requiring an experience based trial and error approach to test them), but all of them impose limitations. Long exposure times and high laser powers can damage the sample, highly confocal settings may worsen detection limits due to significantly weaker signals, and equipping a wide range of laser lines is expensive and may impose further limitations (infrared lasers are not well suited for wet samples and glass slides, for instance).

The second limitation is connected to the potential low specificity and complexity of the spectra. These are partly due to the nature of the technique and partly to the nature of the plant material (particularly xylem). The Raman signal is based on molecular vibrations, with many vibrational modes involving the displacement of only a few atoms. Since these vibrations are characteristic for a set of atoms and bonds (a chemical functional group, such as –O–H and –C=C–), they have a certain energy level. As a consequence, they always appear in a specific range of the spectrum ([4], *see* **Note 3**). However, their exact position within this range depends on several factors (*see* **Note 4**), which is useful for identification in pure compounds (spectral fingerprint matching), but can pose problems in complex mixtures. Xyelm can be considered as a complex chemical matrix, mostly consisting of lignin, cellulose, hemicelluloses, and pectins in various ratios (with other classes of compounds, such as proteins, being present in much lower concentrations, *see* **Note 5**). Since many of these biopolymers are chemically related, they contain very similar functional groups (e.g., –C–H, –C–O, and –O–H) in very similar chemical environments. Consequently, the observed bands are broad, overlapping and seldom diagnostic (i.e., difficult to assign to a particular compound, *see* **Note 6**, Table 1). In addition, composite vibrational modes, involving larger sets of atoms (e.g., ring breathing motions of carbohydrates etc.) can also produce Raman spectral bands, which further complicates analysis. Taken together, these mean that spectra should be interpreted with great care, preferentially using multivariate methods. However, they also mean that the recorded spectra are rich in information, and if proper techniques are used both for recording and for data analysis, Raman microspectroscopy can provide a wealth of knowledge about plant materials, rapidly and non-destructively, in situ, and without external agents at sub-cellular spatial resolution. As a consequence, Raman microspectroscopy has found a wide range of application areas in plant sciences. Excellent and thorough overviews of the field are available with illustrative examples, including the compositional and structural analysis of native or modified (genetically, chemically, or enzymatically) cell walls, studies of structure–function relationships in muro (from roots and seeds to wood and leaves), and investigations of the biosynthesis, development, and deconstruction processes of plant cell walls [5, 6].

A common theme for all these application areas is that they are based on the assignment of spectral features to specific chemical compounds and their structure. Therefore, a short overview is given below for each major cell wall component in this respect, complemented by Table 1 that provides a selection of important Raman bands widely used for band assignments [6–26].

#### 1.1.1 Lignin Amount, Distribution, and Structure

Lignin is unique among the major cell wall biopolymers since it contains aromatic rings and consequently it has the most characteristic spectral bands (*see* **Note 6**, Table 1). These bands can be used to track lignin amounts and distribution within the tissue or cell wall layers. Moreover, since the structure of lignin influences the energy of its functional groups and thereby the band positions in the spectra, this information can be utilized to gain structural insight, including S/G ratio determinations in situ (*see* **Note 7**).

**Table 1**
**Selected Raman spectral bands used for assignments in major cell wall biopolymers [6–26]**

| Cell wall biopolymer | Band position ($cm^{-1}$) | Band assignment | Note |
|---|---|---|---|
| Cellulose | 2890–2900 | –C–H stretch<br>Orientation sensitive | Orientation sensitive |
| Cellulose | 1378 | –CCH, –CHO, and –COH bend<br>Orientation sensitive | Orientation sensitive |
| Cellulose | 1120 | –COC– symmetric stretch (glycosidic) | |
| Cellulose | 1098 | –COC– asymmetric stretch (glycosidic)<br>Orientation sensitive | Orientation sensitive |
| Cellulose | 971 | –CHH<br>Orientation sensitive | Orientation sensitive |
| Cellulose | 880–900 | –COC– skeletal of beta anomers | |
| Cellulose | 380 | –CCC ring bending | |
| Hemicellulose | 2930 | –C–H stretch | |
| Hemicellulose | 1740–1755 | –C=O stretch | |
| Hemicellulose | 1256 | –CH, –COH bend | |
| Hemicellulose | 870–900 | –COC– skeletal of beta anomers | Sensitive to esterification/ etherification of the close carboxyl |
| Hemicellulose | 475–515 | Composite bending vibrations involving the C6 position | |
| Lignin | 1600–1610 | Aromatic skeletal –C=C– vibrations | Used in S/G ratio determination |
| Lignin | 1659–1670 | Aromatic skeletal –C=C– vibrations in coniferyl alcohols, or conjugated to –C=O in coniferylaldehyde | Used in S/G ratio determination |
| Lignin | 1336 | Aryl-OH, aryl-O-Me | Contributions from –CH deformation and –CCH bending |
| Lignin | 1275 | –CH, –COH | Predominantly from G type lignin |
| Pectin | 2945–2950 | –C–H stretch | |
| Pectin | 1740–1755 | –C=O stretch | |
| Pectin | 825–860 | –COC– skeletal of alpha anomers | Sensitive to esterification/ etherification of the close carboxyl |
| Pectin | 816–817 | –COH ring vibrations | |

Band positions are approximate, as they depend on the instrumental setup (including lasers used and spectral resolution) and the chemical matrix of the sample. Please note that bands may not be diagnostic, i.e., they can feature in several compounds (e.g., –C=O band in hemicelluloses and pectins, –C–O–C– band in cellulose and hemicellulose)

#### 1.1.2 Pectin Amount, Distribution and Compositional Changes

Pectins are unique among the carbohydrate based polymers of the cell walls in having α-glycosidic –C–O–C– linkages (whereas cellulose and hemicellulose have β-glycosidic linkages). Therefore, the sharp Raman band around or below 860 $cm^{-1}$, originating from alpha conformations, can be diagnostic of pectins, as the corresponding beta conformations have higher energy, producing a band in the 880–900 $cm^{-1}$ region (Table 1). However, other bands in the 850–870 $cm^{-1}$ region can disturb band assignments and particularly intensity determination. In addition, the exact position of the anomeric band is sensitive to the state of uronic carboxyls, and thus it varies with esterification (acetylation, for instance, shifts the alpha anomeric band to higher wavenumbers, i.e., closer to the beta conformation) and etherification (methylation, for instance, shifts the band to lower wavenumbers, i.e., lowering beta anomers of hemicelluloses closer to the alpha position of pectins). Thus, while they may provide pectin structural information, this can be difficult to extract in a complex chemical matrix containing both hemicelluloses and pectins with mixed structures.

#### 1.1.3 Cellulose Amount, Structure, and (Microfibril) Orientation

Pure cellulose has a set of infrared and Raman active vibrations (Table 1), some of which originate from functional groups in a fixed arrangement (i.e., parallel or perpendicular to the longitudinal axis of the polymer chain/cellulose microfibril). These can be used to estimate the crystallinity of cellulose [27] and by means of polarizers even the direction of the polymer chains (i.e., microfibril angle), both in individual fibers and in tissue sections [5, 6, 28] However, care must be taken to identify band intensity alterations that result from directional effects and not from differences from general cellulose content variation. Thus, it is necessary to monitor all cellulose bands simultaneously, not only the orientation sensitive ones.

#### 1.1.4 Hemicellulose Amount and Structure

Hemicelluloses represent one of the weak areas of Raman (micro) spectroscopy. They are mostly composed of functional groups that are commonly found in pectins and cellulose, and are thus seldom diagnostic (Table 1). Following changes in hemicellulose composition (acetylation etc.) is in theory possible by for example monitoring the position of the –C=O stretch. However, in practice this is often unfeasible, due to the generally weak, broad and unspecific hemicellulose signals in Raman spectroscopy [10], including the –C=O band. Moreover, since hemicelluloses are amorphous polymers, their signals cannot be resolved from the rest of the chemical matrix by means of polarizers and directional changes.

### 1.2 Data Analysis

Since every Raman spectrum recorded contains information about all compounds simultaneously, the entire chemical composition (chemical fingerprint) of the sample is monitored. In addition, microspectroscopic (hyperspectral) data sets also contain positional information, and in case of polarizers even directional information (such as the alignment of cellulose microfibrils). The aim of the data analysis procedure is to extract the maximum amount of information from the data in the most effective way, while avoiding the introduction of artifacts and bias. Spectra usually need to be preprocessed to remove cosmic rays and correct the baseline. Additional preprocessing (including smoothing, noise filtering, and normalization) may be needed, depending on the overall goals and the method of choice for the analysis. In case of image data, visualization of the results in forms of chemical maps is often performed as well, optionally in combination with segmentation (i.e., identification of chemically similar areas in an image). Quantitative information, however, can be limited, especially if normalization is applied, resulting in proportional changes, not absolute amounts (semiquantitative analysis). The key concepts of data analysis steps and methods are explained below.

### 1.2.1 Preprocessing

The first preprocessing step for Raman spectra is the removal of cosmic rays from recorded spectra. The term "cosmic rays" refers to extremely high-energy radiation of various origins (supernovae, active galactic nuclei, etc. [29]), consisting mostly of protons and atomic nuclei that produce a range of secondary particles upon entering Earth's atmosphere and thereby influence electronic devices. In case of Raman instruments, the charge coupled devices (CCDs) used as detectors (*see* **Note 8**) are influenced, producing sharp and often high intensity spikes at random positions in the spectrum. There are several algorithms for removing cosmic rays [30], but none of them can be completely automated, thus requiring user supervision (*see* **Note 9**).

Once cosmic rays are removed, noise in the spectra may need to be lowered. For a single spectrum, the most common way to achieve this is via smoothing [31], while for related datasets (i.e., a series of spectra or image data), multivariate noise filtering approaches provide the best results (including principal component analysis—PCA methods [32–34], wavelet transformation [35], etc.). It is imperative to make sure that this noise filtering/smoothing does not result in any loss of information or introduction of artifacts (by for example over-smoothing by the Savitzky–Golay filter [34], or including too few principal components in the reconstructed data during PCA based noise filtering).

Unlike smoothing or noise filtering, baseline correction is practically a necessity for virtually all Raman spectra, even in case of relatively flat baselines, to correct for the constant background of the CCD. In the case of plant material, however, baselines can be far from linear, having potential contributions from, for example, various levels of fluorescence, etc. While most software packages (including WiRE) offer options for polynomial baseline correction, these are often suboptimal, irreproducible, and labor intensive

[36]. In addition, finding the correct polynomial baseline that would suit all spectra in a dataset (i.e., all pixels in an image of a plant tissue section) can be nearly impossible, since both band positions and baseline shape can change drastically from one spectrum to the next (due to the optical and physical properties varying in the sample, coupled to the chemical compositional differences among cell types, or even cell wall layers [6]). Thus, the ideal baseline correction method should not use a fixed shaped baseline, but individually fit a new baseline for each spectrum in the entire dataset, in a reproducible and automated manner. The method that fulfills all these requirements is called asymmetric least squares fitting [34]. Although it was originally developed to correct chromatographic data [37], it excels at correcting the baselines in vibrational spectra [34].

Further spectral preprocessing steps are optional, depending on the dataset, the aim of the study and the kind of data analysis that follows preprocessing. One of the most common optional preprocessing steps is intensity normalization. It is ideal for removing intensity variation in the dataset due to the properties of the sample and the method itself (out-of-focus effects, partial pixel coverage, hot spots, etc.). Total intensity normalization, for instance, ensures that all pixels have the same total intensity count, while normalization to a certain band can serve as an internal reference calibration (assuming that the band exists in all spectra and its intensity can be determined accurately). As a result of any normalization, the remaining intensity variations represent proportional concentration changes, not absolute amounts. This can facilitate the interpretation of chemical maps, or pixel classification, where categorization is based solely on qualitative information, and is often used prior to univariate analysis (e.g., visualizing single band intensity). However, intensity normalization is generally not needed prior to multivariate image analysis, since those methods (particularly curve resolution techniques) may inherently handle this variation [34]. Similarly, derivatization (another optional preprocessing step) is generally not recommended prior to multivariate analysis, although it can be helpful for quickly visualizing band positional shifts in a purely qualitative way, using univariate techniques.

### *1.2.2 Univariate Data Analysis*

The classical approach to handle hyperspectral datasets (including Raman microspectroscopic images) is by using univariate data analysis. In essence, this means the tracking of a single variable (most commonly the intensity at a single position, or the area (integral) of a single band, or a band intensity ratio) across all spectra to visualize the spatial distribution of this feature in the sample. While this approach has been successfully employed on plant materials [28, 38], it has significant limitations. Firstly, the tracked variable needs to be diagnostic of a certain feature (such as a certain chemical compound), and needs to be known in advance.

These are particularly dangerous assumptions, as they do not consider unknown or unexpected features in the spectra, as the tracked variable is assumed to be dependent only on the correlated feature, ignoring all other influences (band positional shifts, overlaps, noise, etc.). Moreover, they can introduce significant bias and limit the information gained from the data, as no information is gained about the relationship of the tracked variable to all other variables in the spectra. Thus, the use of univariate analysis should be limited to gain a quick overview of the data, including quality checking (i.e., identifying problematic pixels in an image, damage to the sample, or detecting discrepancies between the white light image and the Raman spectroscopic data, due to sample drift or focusing problems, etc.) before multivariate analysis is performed.

#### *1.2.3 Multivariate Analysis*

While multivariate approaches are generally more complex than univariate analysis, they are significantly more powerful as well. They are capable of removing variation not correlated to the question at hand (experimental error, biodiversity, etc.), and they avoid bias by taking into account the entire spectral information, not only selected variables. They are all based on the reduction of the dimensionality of the dataset by finding or generating key variables that can describe the data without loss of information. A plethora of methods exist to achieve this [32, 33], and their detailed description is far beyond the scope of the present chapter. Instead, the basics of the most common multivariate analysis techniques are provided below, and the reader is referred to detailed protocols regarding their practical applications that exist elsewhere [6, 32–34].

Principal component analysis (PCA) is one of the most common unsupervised multivariate methods used in spectroscopy [33]. It is very effective in reducing data dimensionality by generating a set of new variables (principle components) that incorporate all original variables (intensities at all wavenumbers) with various contribution levels. The resulting model describes the original data to a high degree, retaining all relevant information and providing a measure of correlation within the data points (scores) and among the original variables (loadings). A simple PCA analysis is often sufficient for high quality data with significant differences between samples (or pixels within the image). It can also be used to find and filter noise from image datasets.

Another common unsupervised method for multivariate data analysis is cluster analysis. It is based on identifying spectral similarities and grouping (clustering) spectra accordingly. Based on the methods used for determining spectral similarities and classification of the spectra, several versions of cluster analysis exist, including hierarchical cluster analysis, K-means cluster analysis, and fuzzy clustering. Differently from PCA, they can introduce a certain level of supervision by allowing the adjustment of cluster boundary conditions. Cluster analysis is also very effective in reducing the dimensionality of the dataset, since instead of visualizing each band in the spectra, each cluster will be based on the entire spectral profile, providing a quick overview. However, cluster analysis methods can be particularly sensitive to unwanted variations in the dataset, including noise. Thus, they are especially powerful when such variations are already removed, either by preprocessing and/or by other multivariate techniques (such as PCA, or multivariate curve resolution [34]).

Multivariate curve resolution—alternating least squares (MCR-ALS) is a curve resolution method based on the assumption that each spectrum in the data can be described as a liner combination of a set of pure components. The analysis aims to identify these pure components by resolving their spectral profiles and their concentrations (amounts or proportions) in each spectrum of the dataset, through a series of iterative least squares fitting steps (alternating between concentration and spectrum determination) until convergence is achieved. While it is a primarily unsupervised method at its base, a large degree of user control can be introduced via different constraints (limiting the possible solutions of the equations), convergence criteria and methods for initial condition estimations. Its main advantage is the directness of the method, as constraints are based on physical and chemical properties (such as non-negative concentrations) with the results being presented as pure component spectra and concentrations, rather than abstract mathematical representations. This allows the direct observation of the effects of user controlled variables (such as constrains) on the results, and makes the method easy to understand and experimentally fine tune even without chemometrics background. As such, it is a recommended method for plant scientists who want to analyze their spectroscopic data but lack thorough knowledge or experience in the field, with a detailed step-by-step protocol available elsewhere [34].

However, it can happen that unsupervised methods are unable to filter all unwanted variation from the dataset (PCA components plagued by noise, cluster analysis refining areas of a map that are irrelevant, MCR-ALS resolved profiles are not pure enough, etc.). While most methods provide options or at least viable workarounds to compensate for this, it can be beneficial to use a supervised approach instead of trying to force more user control into unsupervised methods. Supervised methods are based on the user predefining certain variables that can be used to direct the analysis. In discriminant analysis (DA), for instance, the user can assign values (Y variables) to certain spectra prior to the analysis. In essence, certain spectra will be classified as belonging to one class, while others will be assigned to another class (supervision via discrimination). The analysis then tries to find the factors that can be used to most effectively differentiate these predefined classes, based on all spectral information available (X variables), while ignoring the variation that does not contribute to class separation (noise). This can be achieved via partial least squares (PLS) fitting and further refined by rotating the results in a manner that facilitates data interpretation and visualization (orthogonal projections to latent structures—discriminant analysis, OPLS-DA [39, 40]). OPLS-DA in particular has been shown to be a powerful method to identify the sources of variation in vibrational spectroscopic data of biological tissue, including plant materials [40].

While undoubtedly powerful, no multivariate method can fully compensate for the potential lack of specificity in Raman spectra. Thus, complementary analytical techniques (e.g., wet chemical analyses, FTIR microspectroscopy, NMR spectroscopy, mass spectrometry, metabolomics, or proteomics) may be required to refine the results or gain more detailed insight into the underlying molecular or genetic processes that result in the observed chemical changes.

## 2 Materials

### 2.1 Chemicals

Since Raman microspectroscopy does not require excessive sample preparation (e.g., no staining or extraction is necessary), the only chemicals required are microscopy oil/water for immersion lenses, $D_2O$ for wet measurements if $H_2O$ interferes (depending on the signals and the lasers applied), and appropriate sample carriers (normal microscopy slides or Raman silent $CaF_2$ windows). Mounting medium for cryotomy may be required for sample preparation.

### 2.2 Instrumentation

The protocol provided was developed on a Renishaw inVia confocal Raman microscope, equipped with five laser lines (457, 488, 514 (tunable $Ar^+$-ion), 633 (HeNe), and 785 nm (diode)) with their maximum power ranging from 15 to 300 mW. The laser was focused using 5–100× magnification objectives, including long working distance, oil and water immersive lenses. Experimental settings are given for this system, but other instrumental setups should operate in analogous fashion, although the software controls are likely to vary somewhat. Thus, parameters given here can be used as initial values, with need of fine-tuning for other instruments and samples.

The sample preparation steps detailed in the protocol refer to a Microm HM 505 E cryostat microtome but are adaptable to other types of cryotomes with minimal alterations.

### 2.3 Software

The measurement steps described in the protocol were performed on standard PCs running Microsoft Windows XP and Windows 7 operating systems, using Renishaw's WiRE Software (Renishaw Plc., UK, http://www.renishaw.com, versions 3.2–3.4, *see* **Note 10**). Other manufacturers provide their own software bundles for instrument operation and we refer the users to their own manuals for finding and adjusting parameters outlined in the present protocol. For data analysis, Renishaw's Cosmic Ray Remover script (part of WiRE's chemometrics package) was used for initial cosmic ray removal and PCA based multivariate noise filtering. Data was exported in ASCII format as simple ".txt" files, which are possible to read by common text processing applications and imported into for example Microsoft Excel. For spectral preprocessing and data analysis, the ".txt" files were processed by a custom-built open source MATLAB script, available free of charge at http://www.kbc.umu.se/english/visp/download-visp/. The script enables saving the data in formats that can be read directly by other software packages for multivariate analysis (e.g., SIMCA-P+, Umetrics AB, Umeå, Sweden).

## 3 Methods

### 3.1 Sample Preparation

While theoretically no sample preparation is required, it is often beneficial to section wood material in the thickness range of 8–20 μm. Sectioning ensures a flat surface, which makes focusing easier, while the given thickness range allows for optimal quality brightfield images to be captured. Thicker, cruder cuts can be easily obtained by hand, but it can be difficult to obtain sufficient quality sections that way. While conventional microcut techniques, including vibratomes may work, cryosectioning often provides the best quality sections and is therefore the method of choice detailed here.

Embedding the sample can help with sectioning (especially for softer materials), but it is generally not recommended, in order to avoid bands from the embedding medium contaminating the spectra. Thus, no embedding steps are included in the protocol. However, if it is deemed unavoidable, PEG (Polyethylene Glycol) embedding has been shown to be the preferred alternative, with protocols provided by Gierlinger et al. ([6], *see* **Note 11**).

The prepared sections can be mounted on normal glass microscopy slides or special, Raman silent microscopy slides made of polished $CaF_2$. Both have advantages and disadvantages, and thus both options are covered in the protocol below. As a general rule, glass microscopy slides are used when the tissue is measured wet, while $CaF_2$ is ideal for dry sections. Wet samples usually hold the anatomy better (drying can induce fractioning, cracks, and curls) and are able to dissipate the heat from the Raman laser, thereby reducing the risk of burning. The presence of water, however can introduce problems, as it (a) limits the choices of laser, excluding the infrared range; (b) can dissolve chemicals and in addition produce a signal in itself that can overlap with other bands (While overlapping water bands can be shifted by using

$D_2O$ instead of $H_2O$, this can induce overlaps at other frequencies instead, and result in the shift of other bands in case of exchange processes); (c) can make sample preparation harder (sections floating, curling/folding, bubbles appearing, etc.). Dry samples, on the other hand, do not have these problems and allow the exact same sections to be processed by complementary techniques (e.g., FTIR microspectroscopy) on the same slides, without any extra steps. They, however, are more vulnerable to the heat generated by the Raman laser and can show distorted anatomical features as a result of drying. Generally, older xylem with thicker cell walls is better suited for dry measurements than younger samples with thinner cell walls in this respect.

#### 3.1.1 Cryosectioning

1. Prior to sectioning, ensure that the sample is prepared to contain only the material to be analyzed (debarked, pith removed, etc.) and, if needed, cut into a small enough size for fitting on the plate for cryosectioning (*see* **Note 12**).
2. Cool the sample by storing it at −80 °C for several hours (typically overnight. *See* **Note 13**).
3. Set the cryostat to working temperatures (typically ca. −10 to −20 °C) using the appropriate up and down buttons on the front panel, and allow the temperature to stabilize.
4. Transfer the sample to the chamber of the cryotome and if needed allow the temperature to stabilize.
5. Apply mounting medium (cryoglue) to the sectioning plate with the sample on it. Make sure that the sample is oriented correctly on the plate and apply enough mounting medium around it to hold it firmly in place, but do not cover too much as it will make orientation adjustments impossible (*see* **Note 14**).
6. Wait until the sample block is frozen solid (typically in the range of a couple minutes).
7. Place Raman transparent $CaF_2$ slides (if dry samples are to be measured; *see* **Note 15**), and/or normal glass microscopy slides (if wet samples are to be measured, and/or if consecutive sections are needed for staining etc.) inside the cryotome chamber.
8. Mount the knife blade into its holder if not already present, secure it with the screws and wait until it is cooled to operational temperature (*see* **Note 16**).
9. Place the sectioning plate with the sample mounted in position for sectioning, and adjust its orientation.
10. Make sure surfaces (including the anti-roll plate) are cleaned from debris from previous sections.
11. Move the sample close to the starting position of the sectioning, either by turning the handwheel or using the built-in

coarse feed function (*see* **Notes 17** and **18**). Make sure that the sample and the knife do not touch during this step (*see* **Notes 19** and **20**).

12. Set the desired section thickness (typically between 8 and 20 μm, *see* **Note 21**) and, if needed, the trimming thickness (**step 14**), using the appropriate control buttons on the cryotome front panel (*see* **Note 22**).
13. Turn the handwheel carefully so that the sample touches the blade of the knife.
14. If trimming is needed (e.g., to reach a deeper layer of the sample quickly), press the trimming button on the front panel slightly (to select pulsed trimming, i.e., advancing with the set trimming value once), or hold it in for about 2 s (to select continuous trimming with the set value until the button is pressed once more). Check the orientation of the sample during trimming, and return to **step 9** to fine-tune if necessary.
15. When the sample is in the correct position (and correct depth if trimming was applied), start the sectioning by turning the handwheel slowly and evenly. Indicators of the sectioning process can be seen on the front panel display, including section count, sum of section thicknesses (in microns), and remaining travel to front end position (in microns).
16. It is critical that sections are of good quality, i.e., not fractured and have smooth, planar surfaces, and it often pays to spend time to finely adjust sectioning, as the spectral results can be greatly improved by improved sample preparation. Therefore, check the quality of the section by placing it on a cold microscopy slide, taking it out from the cryotome, washing it from the cryoglue, air-drying and then observing it under a microscope. Ideal sections have smooth flat surfaces, homogenous thickness (no folding or ripples) and intact anatomy (no fractures or squashed cells) (*see* **Note 23**). Make sure that the orientation of the section is correct. If not, return to **step 9** to adjust.
17. If the sectioning quality is sufficient (*see* **Note 24**), keep sectioning by turning the handwheel slowly and evenly until you have sufficient number of sections (including consecutive ones for parallel staining etc.). Place the sections for Raman microspectroscopy onto cold $CaF_2$ slides (dry sections, *see* **Notes 25** and **26**), and/or on normal glass microscopy slides (consecutive sections, *see* **Note 27**, and/or wet samples).
18. Remove the slides from the cryotome, wash them gently from the mounting medium (*see* **Note 28**).
19. For wet measurements only, place a droplet of water (or $D_2O$) on the prepared section. Apply glue (*see* **Note 29**) on a standard

microscopy cover slip and seal the section with it, making sure the section is not squashed, folded, or otherwise distorted, and there are no bubbles. Let the glue dry.

20. For dry measurements only, place the slides in a desiccator until measurements (typically for at least 24 h to ensure they are dried completely from washing).
21. Once all slides are prepared, remove the remaining sample from the cryotome and clean up.

### 3.2 Single Spot Measurements

Before mapping measurements are started, the system should always be optimized for the sample at hand. This is best done by a series of single spot static scans (*see* **Note 30**), iteratively fine-tuning parameters. Once the user becomes familiar with the technique and the sample, this iterative process will be significantly shorter, as almost optimal starting values can be directly chosen based on experience.

#### 3.2.1 Instrument Setup

**Steps 3–4** and **6** may (partially) be omitted for systems that use a single laser line in a fixed setup. However, for instruments with different laser lines, each step of the setup needs to be repeated every time a new laser line is selected, even if it is the same physical laser tuned to a different frequency.

1. If it is not running already, start WiRE on the computer and wait until it loads (*see* **Note 31**).
2. Switch on the laser to be used and, if required, allow it to warm up (*see* **Note 32**).
3. Check that the correct filter is mounted (*see* **Note 33**). Change filter if needed (*see* **Notes 34** and **35**).
4. If applicable, check that the correct grating is mounted (*see* **Note 36**). Change gratings if needed (*see* **Note 37**).
5. If polarization measurements are to be performed, check whether the correct polarizers are installed (*see* **Note 38**) and change if needed (*see* **Note 39**). Make sure they are installed but not active (i.e., they are not in the optical path) at this stage.
6. Select the laser line and the grating from the dropdown list in WiRE (marked 9 in Fig. 1, *see* **Note 40**).
7. Open the sample chamber door (*see* **Note 41**) and place a flat object (e.g., an empty glass microscopy slide or silica wafer) on the sample tray in the microscope for adjusting laser alignment (*see* **Note 42**).
8. Rotate the lens turret to select the lens to be used for the measurements (*see* **Note 43**). Select the correct magnification in the dropdown list in WiRE (marked 6 in Fig. 1, *see* **Note 44**).
9. In WiRE, click on the “Video and eyepieces” and “Sample illumination” buttons (in the group marked 5 in Fig. 1, *see*

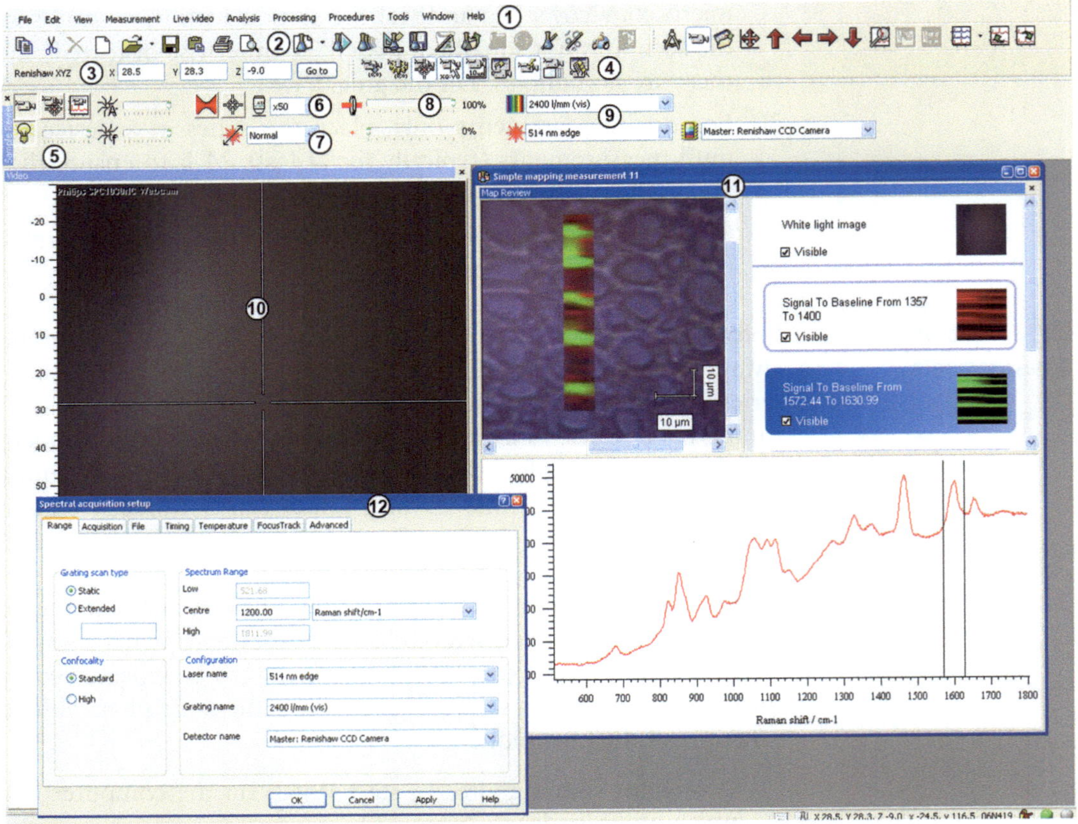


**Fig. 1** The WiRE interface. *Numbered markings* denote the following: (*1*) WiRE menu; (*2*) File (to the *left*) and Measurement (to the *right*) icons; (*3*) *X*, *Y* and *Z* coordinates; (*4*) position and video control icons; (*5*) illumination and live video control icons; (*6*) pinhole and laser shutter icons, lens selection drop-down list; (*7*) polarization selection drop-down list; (*8*) laser intensity control slider; (*9*) gratings (*upper*) and laser line selection (*lower*) drop-down lists; (*10*) live video window with crosshair positioning and scaled axes; (*11*) map review window, with overlaid white light and map display (*left*), layer list (*right*) and spectrum view (*bottom*) with the active map highlighted; (*12*) single spot measurement setup window

**Note 45**). Looking through the eyepiece (*see* **Note 46**), focus on the surface of the flat object (*see* **Note 47**) using either the controls behind the sample tray or the joystick (*see* **Notes 48** and **49**). Make sure to close the sample chamber door when done (*see* **Note 50**) and switch to "Video and Laser" view (in the group marked 5 in Fig. 1, *see* **Note 41**).

10. Fine-tune the focus if needed (*see* **Note 51**), using the joystick.
11. In WiRE, decrease the laser intensity to minimum, using the slider (marked 8 in Fig. 1).
12. If a non-point laser is used (e.g., a diode stripe), apply a pinhole by pressing its button (far left to the lens selection drop-down list marked 6 in Fig. 1).

13. Click on the "Laser shutter" button to open it (immediately left to the lens selection drop-down list marked 6 in Fig. 1). If the laser spot does not appear in the live window (marked 10 in Fig. 1), gradually increase laser intensity by adjusting its slider in WiRE (marked 8 in Fig. 1), until the laser spot becomes visible (*see* **Notes 52** and **53**). Switch off the sample illumination in WiRE (in the group marked 5 in Fig. 1).
14. With the laser spot visible, click on "Tools" > "Manual beam-steer..." in the WiRE menu (marked 1 in Fig. 1) to open the separate beamsteer control window. Click on the control arrows for the left and right beam steer (*see* **Note 54**) until the laser spot is symmetric and is fully centered in the crosshair of the live image panel (marked 10 in Fig. 1).
15. The laser intensity may need to be adjusted during beamsteer control (e.g., decreased for better centering or because the beamsteer adjustment resulted in drastic improvement of laser throughput to the sample; or increased to better observe symmetry etc.). If that is the case, save the beamsteer values, close the beamsteer control window, adjust the laser intensity by its slider (marked 8 in Fig. 1) and return to **step 14** above.
16. Once optimum beamsteer values are found, save them and close the beamsteer control window.
17. Double-check the focus via adjusting the tray level by the joystick and observing the laser spot behavior. Perfect focus should result in a well-defined laser spot with sharp edges (*see* **Note 55**).
18. Return to **step 14** above if the laser is not symmetrical or not centered perfectly after focus adjustment.
19. Close the laser shutter (immediately left to the lens selection drop-down list marked 6 in Fig. 1).
20. Click on "Tools" > "Calibration" > "Quick Calibration" in the WiRE menu (marked 1 in Fig. 1). This starts an automatic calibration procedure, using the built-in Si standard in the podule. Quick calibration takes only a few seconds, during which a status window will be opened, the laser intensity will increase dramatically in the live image window (marked 10 in Fig. 1) and the cross-hair may jump. When the process is finished, the laser exposure is terminated; the crosshair returns to its original (center) position and the status window will display the message that the calibration was successful (*see* **Note 56**).
21. When calibration is finished, the setup is complete and the system is ready for measurements. Press the door release button (*see* **Note 36**) to open the sample compartment and remove the flat object used for setup.

*3.2.2 Measurement Parameter Optimization*

The procedure below describes the steps to be taken when a (type/set of) sample is recorded for the first time, and optimal recording parameters need to be found iteratively. When measurement parameters (such as exposure time and laser power) have been established for a type of sample previously, those values can be used directly instead (particularly in **steps 11–13**), significantly shortening the optimization steps (*see* **Note 57**).

1. Place the sample on the tray in the sample chamber.
2. In WiRE, click on the “Video and eyepiece” and “Sample illumination” buttons (in the group marked 5 in Fig. 1) to activate the camera and the eyepiece (*see* **Note 41**) and enable viewing the sample on the computer screen in the live view window (marked 10 in Fig. 1).
3. Change to the correct magnification lens on the turret and also in WiRE (*see* **Notes 38** and **39**).
4. Adjust illumination by the sliders for light intensity and aperture (in the group marked 5 in Fig. 1, *see* **Notes 58** and **59**).
5. Find a representative position in the sample, using the joystick to move in the *x*/*y* direction (*see* **Notes 41** and **44**). Do not use a position in the sample you intend for mapping later on, but a location that corresponds to it (same type of cell/tissue etc.). This position will be used to identify the correct measurement parameters, and in an unlucky case, it can be accidentally damaged in the process (*see* **Note 60**).
6. Focus on the surface of the sample, using either the controls behind the sample tray or the joystick (*see* **Notes 43** and **44**). Make sure to close the sample chamber door when done (*see* **Note 45**).
7. In WiRE, click on the “Video and Laser” button (in the group marked 5 in Fig. 1) and fine-tune the focus if needed, using the joystick (*see* **Note 47**).
8. Use the laser to determine correct focus by following **steps 11–13** and **17** in Subheading 3.2.1.
9. When optimum focus is found, click on the “Set Origin” button (in the group marked 4 in Fig. 1) to set the *x*, *y*, and *z* coordinates to 0 at the current point (displayed in the respective text boxes in the group marked 3 in Fig. 1).
10. In the WiRE menu (marked 1 in Fig. 1), click on “Measurement > New > Spectral acquisition”. A new dialog window appears (marked 12 in Fig. 1), where measurement parameters can be set.
11. In the “Range” tab of the setup window, select the “Static” option for grating scan type (*see* **Note 25**), and place the center around 1200 $cm^{-1}$ for xylem. The exact number depends

on the laser and grating, to make sure the 840–1770 $cm^{-1}$ range is covered (*see* **Notes 61** and **62**).

12. In the "Acquisition" tab, set the exposure time to the minimum allowed, or to the latest used acceptable value from previous measurements.
13. In the "Acquisition" tab, set the laser power to the minimum allowed or to the latest used acceptable value from previous measurements.
14. Click "Apply" (*see* **Note 63**) and then "OK" to close the window.
15. Click the "Run" button to start the measurement (in the group marked 2 in Fig. 1). When the measurement is finished, the recorded spectra will be shown in a new window.
16. Check the intensity counts on the recorded spectrum (highest value on the *Y* axis) and if there are any bands visible in the spectrum (*see* an example in Fig. 2).
17. If there are no bands but the count is high (several tens or hundreds of thousands, or even saturated), there is likely a fluorescence (or phosphorescence) problem (Fig. 2). This could be caused either by impurities or by the sample itself.

    (a) To investigate if it is caused by impurities, click on the "Video and Laser" button and return to **step 5** to find another representative position. Repeat **steps 6–8** to set the focus in this position, and then proceed from **step 15** (*see* **Note 64**). If the problem persists, it is likely to be caused by the sample itself. Proceed to point b below.

    (b) If the problem is caused by the sample itself, there are several possibilities to counter it, but their chance of success depends on the properties of the sample.

    - Use a high confocal setting. Return to **step 10** and set up a new measurement. In the "Range" tab, under the "Confocality" settings, mark "High" (*see* **Note 65**), then proceed normally from **step 11**.
    - Quench the fluorescence. It is a rather unpredictable method with an increased chance of damaging the sample (especially dry ones and during mapping), and significantly increases measurement times. Thus, use this option restrictively. Return to **step 10** and set up a new measurement. In the "Timing" tab, under the "Sample bleaching" settings, give a bleaching time of 30 s at start, and then proceed normally from **step 11**. Check for sample damage (**step 19** below), and if there is none, repeat this step by gradually increasing the bleaching time, until fluorescence is quenched. If the

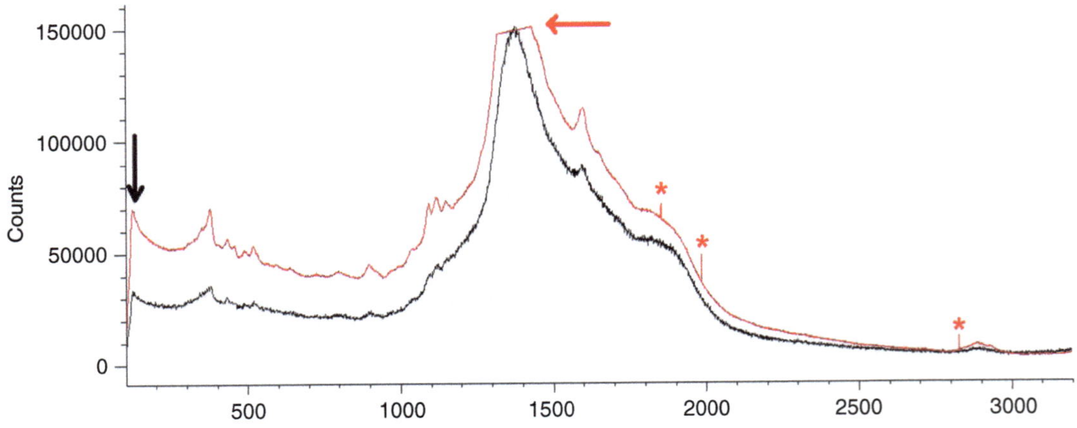


**Fig. 2** Fluorescence problems in the spectra. The *black spectrum* contains low signals, except for the large, broad feature between ca. 1000 and 2000 $cm^{-1}$, originating from fluorescence. Increasing laser power improves signal to noise ratio (*red spectrum*), but fails to quench the fluorescence signal, resulting in saturation (*red arrow* marking the cut band). The cutoff of the edge filter is marked by the *black arrow* (illustrating the lower end of the spectral region that can be studied), and cosmic rays are labeled by *red asterisks*

sample is damaged or impractical recording times are reached before the fluorescence problem disappears, this method cannot be used.

- As a final resort, changing the laser frequency could be the only alternative. If fluorescence problems are caused by a visible laser, lasers in the ultraviolet (UV) or more often in the (near) infrared region may help. Return to Subheading 3.2.1 Setup, and setup a different laser line.

18. If there are no bands in the recorded spectra and the count is low (below (several) thousands, depending on the setup), or the signal to noise ratio is low, the laser power and/or exposure may be increased, provided there is no sample damage.
19. Check the sample for signs of visible damage by clicking on the "Video and Laser" button and adjusting the illumination (**step 4** above).
20. If there is no visible damage, repeat **steps 10–15** but use gradually higher laser power values in **step 13**, until optimum signal to noise ratio is reached and the sample is still unaltered (*see* **Note 66**).
21. If the sample is damaged, but acceptable signal to noise ratio was reached using a lower laser power without damage (*see* **Note 60**), click on the "Video and Laser" button, return to **step 5** to find another representative position and repeat **steps 6–15** using the second highest laser power without damage in **step 13**, to confirm that such laser settings are indeed safe and provide acceptable signal to noise ratio.

22. If maximum laser power (100%) is reached without sample damage or spectrum saturation and the signal to noise ratio still improves, repeat **steps 10–15** but either use gradually longer exposure times in **step 12** until optimum signal to noise ratio is reached (*see* **Note 60**), or double-check optimal focus by adjusting focal depth. To do this, type a different number for the *z* coordinate value in WiRE (in the group marked 3 in Fig. 1), going a few microns deeper into (– numbers) or higher above (+ numbers) the sample than the original focus position (*see* **Note 67**), click the "Go to" button and repeat **steps 10–15** until optimum signal to noise is reached and the sample still remains unaltered.
23. When parameters for obtaining optimum signal to noise ratio without sample damage are found, note the laser power and the exposure time used, together with high confocality or bleaching time values if used. If the measurements do not need polarizers or high confocality (or confocality was already set to high in **step 17a** due to fluorescence), the setup is hereby completed, proceed to **step 31**.
24. If confocal measurements are to be performed, but they were not set up in **step 17** (in connection with fluorescence problems), return to **step 10** and set up a new measurement. In the "Range" tab, under the "Confocality" settings, mark "High" then proceed from **step 11**, using the parameters noted in **step 23** (*see* **Note 65**).
25. If the signal to noise ratio is suboptimal in the confocal measurement, return to **step 20** and use gradually higher laser powers, or if already at maximum laser power, return to **step 22** and use gradually higher exposure times until optimum signal to noise ratio is reached and the sample is still unaltered (*see* **Note 60**).
26. When parameters for obtaining optimum signal to noise ratio without sample damage are found, note the laser power and the exposure time used, including bleaching time values if used (**step 17b**). If the measurements do not need polarizers, the setup is hereby completed, proceed to **step 31**.
27. If orientation sensitive data is to be recorded, a polarizer needs to be applied (**step 4** in Subheading 3.2.1, *see* **Note 68**), preferentially together with a polarization stage (*see* **Note 69**). Open the instrument hatch and apply the required polarizer by folding it into the optical path (*see* **Note 70**). Using a polarizer means that functional groups aligned with the polarization direction (matching either the *x* or the *y* axis, depending on the polarizer/half plate used) will interact with the light (produce the highest intensity signal), while functional groups perpendicular to it will be silent (produce no signal) (*see* **Note 71**).

28. In WiRE, select the polarization direction from the drop-down menu (marked 7 in Fig. 1) to match the polarizer applied (*see* **Note 72**). "Normal" means 0°, "Orthogonal" means 90°, "Circular" means 45° (i.e., diagonal to the cross-hair in the live video window).
29. Return to **step 10** and set up a new measurement. In the "Advanced" tab, in the "Input polarization" field, select the polarization direction according to **step 28**. Proceed from **step 11** using the parameters noted in **step 23** (normal confocality measurements) or **step 26** (high confocality measurements).
30. To observe if any bands are sensitive to directional changes, return to **step 29** and change the polarization direction (*see* **Note 73**). Compare the resulting spectra to find bands with altered intensity. If there are none, return to **step 5** to select a different measurement spot where ordered structures are suspected, and repeat **steps 29** and **30**.
31. If no mapping is required, only single spot measurements at various positions, return to **step 5**, find the position intended for measurement, and repeat **steps 6–15** using the parameters noted in **step 23** (normal confocality measurements) or the parameters noted in **step 26** (high confocality measurements), giving the correct polarization direction in the "Advanced" tab, in the "Input polarization" field, in case of orientation sensitive measurements (*see* **Note 74**). Click on the "Run" button in the WiRE (in the group marked 2 in Fig. 1) to record the spectrum.
32. Save the recorded spectrum via the "File" menu in WiRE, giving it a representative name (*see* **Note 75**).
33. Repeat **steps 31** and **32** for all samples, positions and polarization orientations.

### 3.3 Raster Scan Mapping

Before mapping can take place, the instrument setup (Subheading 3.2.1) needs to be completed and optimum measurement parameters established through single spot measurements (Subheading 3.2.2, *see* **Note 57**).

1. Perform **steps 1–3** in Subheading 3.2.2.
2. Perform **step 4** in Subheading 3.2.2, but keep in mind that WiRE takes a visible snapshot over the measurement area at setup, and NOT at the start of the measurement. Thus adjust illumination at this stage to obtain the nicest white light image.
3. Perform **step 5** in Subheading 3.2.2. When choosing the mapping location, however, keep in mind that the sample needs to be stable during the measurement (this is seldom a problem for xylem samples) and at least 100 points should preferentially be

collected to facilitate multivariate analyses. Thus, find a region in the sample where a large enough area has good optical and physical qualities (e.g., in focus, flat, and anatomically sound)

4. Perform **steps 6–9** in Subheading 3.2.2.
5. In the WiRE menu (marked 1 in Fig. 1), click on "Measurement > New > Map image acquisition". A new window appears, where the measurement area needs to be defined.
6. Select the type of area to be mapped to be filled rectangle (*see* **Note 76**). Click and drag on the white light image panel to define the region. WiRE automatically adds the step size (in microns) and calculates the number of points to be measured (*see* **Note 77**). Change both the *X* and *Y* step sizes to the desired values (*see* **Note 78**), depending on the accuracy of the stage (*see* **Note 79**), the spot size of the laser (*see* **Note 80**) and purpose of the measurement. For the most effective multivariate noise filtering, make sure to use at least 100 points in total/image. WiRE will automatically recalculate the total number of points and updates the display based on the entered step sizes (*see* **Note 81**). Set the area coverage (i.e., direction of the tray movement) to either raster (preferred) or snake (*see* **Note 82**). Once the area, scan direction and step sizes are defined, click "OK". If the measurement area is incorrectly drawn and a new one needs to be created, click "Cancel" and return to **step 5** (*see* **Note 83**).
7. Once the map image area selection window closes, the measurement setup window appears. It is almost identical to the one used for single spot measurements (marked 12 in Fig. 1), except that it contains one more tab, entitled "Area Setup". This tab is identical to the one used in the previous step, allowing for last minute changes to step size and *X* and *Y* starting/ending coordinates (*see* **Notes 84** and **85**).
8. Perform **steps 11–13** in Subheading 3.2.2, using the measurement parameters (spectral region and scan mode, laser intensity, and exposure time) as determined by single spot measurements in Subheading 3.2.2.
9. Differently from single spot measurements, it is recommended to specify a filename for the image data already at setup in the "File" tab (*see* **Note 86**). This way, the data is automatically saved at the end of the recording (*see* **Note 68**).
10. Optionally, a "dummy" integration can be specified, which will be performed "on the fly" to show the progress of the measurement visually, in addition to the standard counter. To do this, click on the "New" button in the "Live Imaging" panel in the "Acquisition" tab. In the new map review setup window that appears, select either "Intensity at position" or "Band area" in the drop-down list and click on the "Create" button.

Specify the band to be mapped by its wavenumber(s) and click "OK" (*see* **Note 87**). Close the map review setup window.

11. Perform **step 14** in Subheading 3.2.2.
12. The measurement is now defined and placed in the measurement queue (*see* **Note 88**). Click the "Run" button to start the measurement. As the measurement progresses, spectra from the last recorded pixel is shown next to the white light image with the measurement grid visible. On this grid, each recorded point will be marked by the intensity of the "dummy" integral if one was selected in **step 10**, visualizing the progress (*see* **Note 89**). A counter is also shown under the recorded spectra, with an estimation of the remaining measurement times (*see* **Notes 90**).
13. When the measurement is done, close its window (*see* **Note 91**) and unload the data.

### *3.4 Data Analysis*

While WiRE is excellent in handling the instrument and performing measurements, the basic package is far from sufficient for data analysis purposes. An optional chemometrics package can (and should) be purchased, which contains excellent tools for cosmic ray removal and multivariate noise filtering for image datasets (not for single spectra). However, it has no additional tools for standard spectral manipulation (baseline correction, intensity normalization, etc.), which are left suboptimal in WiRE. Thus, such tasks are best performed outside of WiRE, and consequently data needs to be exported. This section provides the protocol to manipulate the images to eliminate experimental interferences (noise and cosmic rays) and to export the spectral data (both images and single spectra). Protocols are available elsewhere for multivariate curve resolution—alternating least squares analysis of Raman spectral data [34] and other univariate and multivariate methods [6], which should be consulted for notes and details. Thus, unlike the rest of the protocols in this chapter, only the key conceptual steps are listed in Subheading 3.4.2, with clear indications of significant branching points from where the data can be taken to different types of analysis.

#### *3.4.1 Spectra Manipulations in WiRE and Export*

Method 1 (set of single spectra):

1. In the WiRE menu (marked 1 in Fig. 1), choose "File" > "Open in new window…" (*see* **Note 92**) and select the files to open (*see* **Note 93**). Since there are no good ways of removing cosmic rays from independent single spectra, these spectra should already be free from them (*see* **step 31** in Subheading 3.2.2 and **Note 74**) and thus only require exporting.
2. If not visible, open the Navigator window clicking on "View" > "Navigator Window" in the WiRE menu (marked 1 in Fig. 1).

3. In the "Data" tab of the Navigator Window (*see* **Note 94**), mark the spectrum to be exported (*see* **Note 95**), right-click it and select "Save As..." from the opening context sensitive menu.
4. In the opening standard Windows dialog, specify the folder and the filename (*see* **Note 96**). Select "Text files (*.txt)" in the "Save as type:" drop-down menu (*see* **Notes 97–99**).
5. Repeat **steps 3** and **4** for all spectra.
6. Start Matlab.
7. Right-click in the "Workspace" panel and chose "New" from the opening context sensitive menu. A new variable is created. Give it the name "Data".
8. Repeat **step 7**, but this time give the new variable the name "Filenames".
9. Double-click the new variable "Data" to activate the "Variables" editor panel.
10. Double-click the new variable "Filenames" as well, to load it as a new tab in the "Variables" editor panel.
11. Open each of the saved single spectra ".txt" file (in for example Notepad) and copy–paste their data one by one into the "Data" variable in Matlab. Make sure that the first column contains the wavenumbers, thereafter each column contains one spectrum (intensity values), i.e., the wavenumber column only needs to be included once, NOT from each ".txt" files.
12. Add the name of spectrum (the filename of the ".txt" file) to the "Filenames" variable.
13. Save the "Data" variable as a ".mat" file by right-clicking on its name in the "Workspace" panel and selecting "Save as" from the opening context sensitive menu. Give it a representative name (*see* **Note 100**).
14. Repeat **step 13** for the "Filenames" variable, making sure that it is saved with a representative name, which is not identical to the filename where the "Data" variable was saved (*see* **Note 101**).

Method 2 (image data):

1. Start the Cosmic Ray Remover script (standalone application in the chemometrics package of WiRE).
2. Choose "File" > "Open ..." and select a single image file to open, using the standard Windows dialog.
3. In the "Cosmic ray removal options" box, click on the "Go" button next to the "Nearest neighbour method (recommended)" (*see* **Note 102**). The image data will be loaded. On the left panel, red boxes mark pixels where the software identified potential cosmic rays. On the right, the spectrum from the

highlighted pixel is shown in blue. In red, the neighboring pixel used for cosmic ray estimated is also shown, and in green the resulting spectrum after cosmic ray removal.

4. If the cosmic ray is correctly identified, click on "Accept". The software automatically cycles to the next cosmic ray in the active spectrum (if there are more than one) or the next spectrum with potential cosmic rays identified (if there are no more cosmic rays in the current spectrum) and updates the display (*see* **Note 103**).
5. If the identified feature is not a cosmic ray (*see* **Note 104**), click on ">>" to cycle to the next cosmic ray in the active spectrum (if there are more than one) or the next spectrum with potential cosmic rays identified (if there are no more cosmic rays in the current spectrum). The display is updated automatically (*see* **Note 100**). If needed, use the "<<" button to return to the previous spectrum/cosmic ray.
6. When all potential cosmic rays have been assessed this way, choose a new, lower offset value in the "Parameters" box to the right (*see* **Note 105**) and click the "Find Rays!" button.
7. If there are cosmic rays found at this offset value, return to **step 4**.
8. If there are no cosmic rays found, return to **step 6** and chose a lower offset value.
9. When the software does not allow further lowering of the offset value, or only noise is picked up as cosmic ray by the software, click on the "Done" button to return to the main interface.
10. Click on the "Noise filter" button (*see* **Note 100**). A new window appears, containing an image map based on the scores of the first principal component (i.e., the largest variation in the dataset), and the corresponding loadings plot. The variance explained by PC1 is shown and a plot visualizing the correlation of all principal components loadings is displayed in the bottom.
11. Click on the "Next" button to display the next principal component scores and loadings and the variance explained by this principal component.
12. Repeat **step 11** until the scores plot represents a more or less random image (no identifiable anatomical or image features, *see* **Note 106**) and the loadings plot contains mostly noise (no identifiable spectral features such as bands or background). This usually coincides with (but can also precede) low "variance explained" values for the component (*see* **Note 107**). Note the number of component at which this stage is reached. The final number of components used will be one less than this number (*see* **Note 108**).

13. Enter the final number of component noted in **step 12** in the "Number of (spectral) components" text box and hit "Enter" on the keyboard (*see* **Note 109**).
14. Click the "Noise filter" button and wait until the calculation is finished (*see* **Note 110**).
15. Click the "Exit" button.
16. In the "File" menu, click "Save as..." and in the opening standard Windows dialog, specify the folder and the filename (*see* **Notes 96** and **111**).
17. Exit the Cosmic Ray Removal script by closing its window or by choosing "Exit" in the "File" menu.
18. In the WiRE menu (marked 1 in Fig. 1), choose "File" > "Open in new window..." (*see* **Note 92**) and select a single image file to open (*see* **Note 112**).
19. Visualization of band intensities (univariate mapping) is performed by selecting "Analysis" > "Mapping Review..." and specifying the type of intensity evaluation (band area, band intensity at a fixed point, etc.) and the wavenumber (range) over which the evaluation should take place (*see* **Note 113**). By default, if a dummy integral was specified during measurement setup (**step 10** in Subheading 3.3), it is already shown in the list. However, several maps can be created this way at once, and visualized together. Each maps becomes a new layer in the map review window (marked 11 in Fig. 1, *see* **Note 114**).
20. The image data file will be shown in a new window (marked 11 in Fig. 1), using the white light image captured during measurement setup (**step 7** in Subheading 3.3), together with a band intensity map overlay in the left pane. Immediately to the right, different layers of the map review are showed (including the white light image and all integrals selected in **step 19**). In the bottom pane, the first spectrum of the data set (upper left corner pixel in the image data) is shown.
21. If needed, the map review can be customized by adjusting the layers (*see* **Note 115**). Right click on the layer to be adjusted in the list in the right pane, and chose "LUT control" from the appearing context sensitive menu. Change the color by the drop-down menu, set transparency below a certain value limit by ticking the relevant checkbox, and adjust the range (contrast) by dragging the data and display markers to custom levels (or automatically by clicking on the "5–95 %" level button), as preferred.
22. If needed, save the map review image (to be used as reference in future analysis, for instance), by right-clicking in the map review pane (marked 11 in Fig. 1) and choosing "Save to..." in the opening context sensitive menu.

23. Optionally: save the modified WiRE data file via selecting "Save" in the "File" menu to save the generated mapping reviews from **steps 19–21**. This way, all maps are available next time the file is loaded in WiRE for selection (**step 19**).
24. In the "File" menu, click "Save as…" and in the opening standard Windows dialog, specify the folder and the filename (*see* **Note 96**). Select "Text files (*.txt)" in the "Save as type:" drop-down menu (*see* **Notes 97**, **98** and **116**).

#### *3.4.2 Spectra Processing in Matlab*

Before start, make sure you have downloaded the latest version of the free, open source Matlab script from http://www.kbc.umu.se/english/visp/download-visp/, unzipped it and that both the .m and the .fig files are located in the same folder (*see* **Note 101**).

1. Start Matlab.
2. In the "Current Folder" panel, navigate to the folder that contains the downloaded script. Locate the ".m" file, right-click on it and select "run" in the opening context sensitive menu (*see* **Note 117**).
3. Load the ".mat" file that contains the compiled single spectra (generated in **step 13** in Method 1 in Subheading 3.4.1, *see* **Note 118**) or the ".txt" file that contains a single image exported form WiRE (generated in **step 24** in Method 2 in Subheading 3.4.1, *see* **Note 119**).
4. The graphical user interface is populated by plots of the data (*see* **Notes 120** and **121**).
5. Perform spectrum trimming (cutting) if necessary (*see* **Note 110**) using the appropriate controls of the graphical user interface.
6. Perform spectrum preprocessing by
   (a) Correcting the baseline using asymmetrical least squares fitting, via adjusting the lambda and $p$ values. Details are given in Ref. 34.
   (b) Optionally, perform Savitzky-Golay smoothing by adjusting the order and frame rate parameters. Details are given in Ref. 34.
   (c) Optionally, perform normalization.
7. Branching (exit) point: Save the preprocessed data in ".mat" format, using the button dedicated for this task in the graphical user interface. The data is preprocessed and can be taken to other kinds of analysis now (e.g., imported into SIMCA-P, *see* **Note 122**).
8. Click on the "Pre-Treated for MCR" button to send the preprocessing data for multivariate curve resolution—alternating least squares (MCR-ALS) analysis.

9. Choose MCR-ALS parameters, starting with the number of components (either manually supplying this number, or choosing it based on the singular value decomposition plot), the number of iterations and the convergence criteria (*see* **Note 123**). Optionally, choose the constraints as needed in the graphical user interface.
10. Perform MCR-ALS by clicking on the dedicated button in the graphical user interface.
11. Refine/test MCR-ALS parameters by returning to **step 9** and choosing different values, then performing **step 10**. Compare the results and repeat until optimum parameters are found.
12. Branching (exit point): Save the MCR-ALS resolved spectral and concentration profiles using the buttons dedicated for these tasks in the graphical user interface. The resulting ".mat" files can be used as input for standalone scripts or software packages for visualization or further processing (super resolution, segmentation, cluster analysis, etc.)
13. Visualize the results via the "Show Component Maps" button for evaluation/validation (*see* **Note 124**).
14. Optionally, compare the resolved spectral profiles to reference spectra, using the controls in the "Reference Matching" part of the graphical user interface (*see* **Note 125**). The results are automatically displayed in a table format, and visual evaluation provided as an option.
15. Optionally, perform segmentation based on k-means clustering. Supply the number of cluster either manually, or determine it automatically via silhouette clustering.
16. Branching (exit) point: Save the segmentation results by clicking on the dedicated button in the graphical user interface. The saved ".mat" file of the segmentation result consists of two variables: "IDX" (which stores the cluster membership of each spectrum) and "Centr" (which contains the centroid profiles, describing the contribution of each MCR-ALS resolved component to each cluster). The IDX variable can be used to for example supply *Y*-variables (class identifiers) for discriminant analysis in other software packages, such as SIMCA-P.
17. When the analysis is finished, close the window of the graphical user interface, together with all other open figure windows to finish the analysis (*see* **Note 126**).
18. After the main interface window is closed, type "clear all" in the Matlab Command Window and hit "Enter" to clear all variables from the memory (*see* **Note 127**). Type "clc" and hit "Enter" to clear the Command Window.

## 4 Notes

1. Only vibrations that involve a change in the polarization potential will produce Raman active vibrations. Thus, pure monoatomic and diatomic gases (e.g., He, Ar, Ne, $H_2$, $N_2$, and $O_2$) are Raman silent.
2. Peaks are called bands in vibrational spectroscopy terminology.
3. These are called characteristic group frequencies and are traditionally given in charts for a set of compounds. These charts do not only display the wavenumber range in which a functional group can produce a band, but also show information about band shape (narrow, broad, shoulder) and intensity (weak, medium, strong).
4. The factors include bond order (single, double, triple), the (reduced) mass of the atoms involved in the vibration, and bond strength. The effect of bond strength on the characteristic frequencies means that inter- and intramolecular forces will also affect the band position. An interesting example is the amide I vibration, the position of which is sensitive to protein conformational changes. The effect of the (reduced) mass on band positions can be utilized in isotope studies, either to identify the origin of a band, or to shift a band to a different position to avoid overlaps (using $D_2O$ instead of $H_2O$, for instance, to move the water bands away from the amide I band of proteins).
5. Water could also be present to various degrees. While normally it does not pose as severe problems for Raman spectroscopy as it does for infrared spectroscopy, it can be useful to dry the samples, especially when (near) infrared lasers are used.
6. The most notable exception is perhaps the aromatic –C=C– bond. Since normally lignin is the only aromatic compound that is present in xylem samples in significant amounts, bands appearing around 1600 $cm^{-1}$ are often considered diagnostic of lignin. Note, however, that this is only true if there are no other vibrations producing a band around 1600 $cm^{-1}$, and aromatic –C=C– bonds can only originate from lignin in the sample. In other words, there are no lignin bands, only aromatic –C=C– vibrations that can originate from lignin with higher or lower likelihood.
7. Strictly speaking, the lignin bands are not S- or G-lignin specific, but simply inform about the energy levels of the –C=C– skeletal vibrations of the aromatic ring. Since different levels of substitutions on these aromatic rings result in different energies, they can be indirectly connected to S and G lignin structures or different levels of cross-linking in the lignin structure (Table 1).

8. Traditional silicon based CCDs limit the laser lines to below 850 nm, due to the drop off in their sensitivity beyond this wavelength. Novel multichannel array detectors (using InGaAs based materials instead of silicon) can expand the range well into the near infrared region, allowing the use of longer wavelength lasers (such as 1064 nm, common in biological studies). While these lasers and their setups (including the detectors) are more expensive, they often eliminate common fluorescence problems, albeit at the cost of increasing water sensitivity.
9. The identification of cosmic rays can be based either on the width of features, as cosmic rays are often (but not always) much sharper than normal spectral bands, especially in case of biological tissue. Alternatively, since cosmic rays are extremely unlikely to hit at the same positions twice in a row, they can be identified by analyzing neighboring spectra, either in space (i.e., the neighboring pixels in an image) or in time (i.e., the next spectrum recorded in a series, or the next scan when using multiple scans for the same spectrum). However, none of these methods is failproof, and it can be particularly hard for an algorithm to find (especially lower intensity spikes of) cosmic rays if the neighboring spectra are markedly different.
10. With proper licenses, WiRE can be installed on standalone computers, not connected to the Raman instrument. However, it requires Microsoft Windows XP or newer.
11. Care must be taken not to smear the embedding medium over the sample during sectioning. In addition, the spectrum of the pure embedding medium should be recorded separately and used to check for traces of the embedding medium in all sample spectra, to ensure it causes no interference with the analysis.
12. Section holder plates are typically 3–4 cm in diameter and are at 28 mm distance from the knife blade in their far back position. Thus samples need to be smaller than these dimensions to be fitted for sectioning, allowing room for the mounting medium to secure them in place.
13. Faster (10–30 min) liquid nitrogen cooling can be applied, and various sample cooling accessories may also be available, but they are less practical for large number of samples or for long-term storage.
14. The mounting medium has a high viscosity and can freeze quickly, making orientation adjustments difficult after it surrounds the sample.
15. $CaF_2$ slides are extremely brittle and expensive. Handle them with care.
16. The blades are extremely sharp! There is a handwheel break function implemented in the cryotome with a pin on the rim

of the handwheel. This allows for locking the handwheel and thus the knife. Use this always when not performing the sectioning movement itself, to limit the dangers of working around the blade.

17. For short distance movements, press the coarse feed button only slightly. If the button is pressed and held for ca. 2 s, the sample holder moves automatically to the back end position. This automated movement can be stopped at any time by slightly pressing the button again.
18. For safety reasons, flip the anti-roll plate onto the blade before activating the coarse feed function, to avoid potential splinters.
19. It is particularly important to carefully monitor the knife and sample position during the coarse feed function, which involves rapid movements.
20. Some crytome setups may be equipped with automatic approach systems. For their operational routines including working temperature ranges, consult their manual.
21. The thickness should be varied to match the sample properties. Thinner cell walls (e.g., in younger plants or in early wood) may require thicker sections to hold together.
22. Depending on the type of the cryotome, only certain step sizes may be allowed in selecting the thickness. Graduation may also depend on the thickness range selected.
23. Suboptimal sections are often the results of a blunt blade, inadequate section thickness for the sample (often too thin), or uneven, jerky movement of the handwheel.
24. The first few sections of a series often need to be discarded, even if trimming was not selected.
25. The $CaF_2$ slides have no frosting, but avoid placing section too close to the edges nonetheless, as the slide holder accessories on the Raman microscopy stage will cover those areas.
26. Since $CaF_2$ slides are expensive, it is highly recommended to place several sections on the same slide. However, keep good control over their identity and location, by either marking the $CaF_2$ slides or the corresponding glass microscopy slides in a clear, unambiguous manner. Permanent ink marking is possible on $CaF_2$ slides but should not be done excessively, and only on areas not subject to measurements, to avoid signal from the ink and facilitate easy removal if the slides need to be reused.
27. It is often recommended to mount several sections on the same carrier to be able to select the best one for measurements. Consecutive sections to be saved for staining and inspection under light microscopy should be saved separately. Never stain

or otherwise directly label the sections that are to be measured by Raman microspectroscopy, as the dye(s)/label(s) will influence the spectrum.

28. While the $CaF_2$ slides are water-soluble, their solubility is very low and thus they can be washed with cold water. However, avoid submerging and storing them in warm water for extended periods of time.
29. Strange as it may sound, regular nail polish works perfectly for as a glue.
30. Static measurements allow the user to set only the center of the spectral region, with the lower and upper limits automatically calculated by the software, based on the laser and gratings selected. Thus, the spectral range will be fixed and can only be shifted, not expanded. This can be a considerable limitation if bands are too far apart to be covered within the predefined range. However, static mode measurements are significantly faster than extended mode measurements, and are thus ideal for determining measurement conditions and for rapid measurements (if all important bands can be covered in the spectral range). Extended mode measurements, on the other hand, allow for any spectral range to be set, but at the expense of significantly longer acquisition times.
31. If the system was shut down improperly (e.g., due to a power cut), WiRE may require the motors to be referenced at start-up. This is an automated process and in most cases only requires the user to accept the steps suggested by the software.
32. Some lasers have external cooling systems with their own switches. If that is the case, make sure to switch them on too. Follow the instructions of the manufacturer.
33. The dual passband of notch filters allows users to measure Stokes and Anti-Stokes bands without changing filters. Edge filters, on the other hand, allow light to pass on one side of the Rayleigh line only.
34. Each laser line has its own filter, even if the same physical laser is tuned to different frequencies.
35. It may physically be impossible to change a filter if the filter wheel is in the wrong position from a previous measurement. In that case, change the laser line in the software (marked 9 in Fig. 1) to trigger the filter wheel rotation, allowing access to that position.
36. Although it is not a general feature of Raman instruments, the current setup used a fixed grating for each laser frequency that needed manual changing. In addition, gratings may need to be installed in pairs. Follow the instrument manufacturer's guidelines.

37. Be aware that gratings are extremely sensitive and handle them with extreme care. If dust is noticed, use pressured air to blow clean. Do not use wipes or solvents; do not leave fingerprints or scratches.
38. Each laser line has its own polarizer, even if the same physical laser is tuned to different frequencies. They may be installed in pairs, including a half-plate, to allow switching the polarization direction 90°.
39. Changing the polarizer may require removing a filter from the filter wheel. This may physically be impossible if the filter wheel is in the wrong position. In that case, change the laser line in the software (marked 9 in Fig. 1) to trigger the filter wheel rotation, allowing access to that position.
40. If fixed gratings are used for each laser line, a warning message may be shown when selecting a laser/grating pair that is not matching. Since only one of the drop-down menus (either the grating or the laser, marked 9 in Fig. 1) can be changed at a time, this message may be inevitable when both need to be changed. However, it is only a warning, not an error.
41. The sample chamber has a door release button. While the door can also be opened without this button (immediately activating a built-in safety interlock to avoid laser exposure), it is recommended to use it because it automatically switches on the sample white light illumination (convenience) and, more importantly, it saves the magnetic lock from dislocation (increasing lifetime and avoiding the accidental activation of the safety interlock which would interrupt the measurements).
42. Do not use any real sample for this, because unintentionally high laser exposure may alter the sample and it may be difficult to find a place for perfect focus (due to uneven surfaces etc.).
43. Lower magnification lenses are better for orientation within the sample and for focus adjustments initially, but once the correct position and focus are found, change to the magnification that is intended to be used later in the mapping measurements.
44. WiRE has no information from the turret as to which lens is selected, thus it has to be set manually by the corresponding drop-down list (marked 6 in Fig. 1). Incorrectly set magnification results in erroneous $X$ and $Y$ step sizes in the software, which lead to wrong coordinate readouts, wrong tray movement, and wrong scale bar sizes on the images.
45. For safety reasons, the eyepiece is inactive until the live video image is activated in WiRE, i.e., the microscope cannot be used without the software running. The "Video and eyepiece" button will activate the eyepiece, while the "Video and laser"

button will switch the eyepiece off but keep the live video active on the computer screen, allowing for the laser shutter to be opened.

46. The eyepiece has a larger viewfield than the image displayed in the live image window of WiRE (marked 10 in Fig. 1), making orientation in the sample easier.
47. It is easier to focus on clear features, such as the frosting on the side of glass microscopy slides, or scratches, dust speckles, and permanent ink markings.
48. The sample tray is mounted on a rail in the back of the sample chamber, and thus can only move up and down within a certain range (the length of the rail). If the focus is outside of the current range, the tray has to be manually repositioned on the rail. Make sure you hold the tray firmly with one hand and loosen the screw that holds the tray with your other hand. Then move the tray up or down as required, and secure it again in its new location by fastening the screw.
49. If it is not used for a period of time, the joystick is automatically de-activated. Press any button to activate it. As a general rule, use the button that corresponds to the lens magnification, in order to select the correct speed for tray movement (higher magnification lenses require finer tray movement).
50. The laser cannot be activated on the sample with the door open, due to a safety interlock switch (to avoid accidental exposure).
51. The focus through the eyepiece may not match the one on the computer screen. If that is the case, use the computer screen for focusing because this image is the one saved for the measurements and because when the focus is fine-tuned with the laser, the eyepiece will be deactivated and only the computer screen can be used.
52. The gradual increase of the laser intensity from a low level should ensure that the camera does not get saturated. The most common effect of saturation is only a hue change in the image, as the camera tries to compensate. This may be inconvenient, but it is harmless in a sense that it is impossible to damage the camera this way. The sample can be damaged, however, and this is why no sample is used at the setup stage.
53. Certain lasers might have their own physical power level control. Keep these at a fixed position (i.e., at a fixed laser output power), while adjusting the laser intensity in WiRE.
54. The left beamsteer is responsible for the positioning in the *xy* plane, while the right beamsteer controls the angle of incidence. However, they do affect each other to a certain degree, so repeatedly switching between them is necessary until the optimum is found.

55. During setup, a flat object is used, which facilitates correct focusing. However, correct focus may be difficult to observe on the sample. In that case, moving the sample slightly in the *xy* direction may be helpful.
56. In rare cases, a large offset must be applied during calibration. The status window will notify the user about this via a warning message and will recommend a second run of quick calibration. If that happens, simply repeat this step.
57. Since Raman measurements are extremely customizable, only those options are listed that specifically require input. All other options can be left at their default setting.
58. Higher magnification lenses and thicker, darker samples generally require more light.
59. Some systems may have a separate light source underneath the sample (transmission light), which can be used when more light is needed. However, this extra light source is often controlled manually (including switching it on/off, setting its aperture and intensity). Thus, if the extra transmission light is used for sample viewing, care must be taken to switch it off manually before starting the measurement! Be aware that WiRE automatically switches off the lights above the sample tray and disables the camera and eyepiece, so there is no indication in the live view window (marked 10 in Fig. 1) or elsewhere in WiRE, whether the extra light is on or off. Leaving it on will often result in large counts but low signal, as the white light provides a high background.
60. Too high laser levels, in particular, can burn dry sections during measurement parameter optimization.
61. Since xylem primarily consists of cellulose, hemicelluloses, lignins, and pectins, this spectral range ensures that characteristic bands from these major classes of biopolymers are included and is therefore generally suitable for measurement setup purposes. However, if the measurements aim to specifically monitor bands outside this region, centering around their position may be beneficial instead.
62. If the entire range cannot be covered, aim to include the lower end for samples rich in carbohydrates and the higher end for samples rich in lignin during setup. During sample measurements, however, choose extended scans if a wider spectral range must be monitored.
63. It can be worth double-checking whether the correct spectral region is kept in the "Range" tab. Due to a software glitch, the spectral region may occasionally revert to its default settings.
64. As long as the resulting spectrum window is not closed, there is no need to set up the measurement again. WiRE will

record the new spectrum with the exact same parameters as the ones used for the spectrum in the active window. A warning dialog may be presented to avoid accidental overwriting of existing data.

65. The high confocality setting uses a pinhole to eliminate light from out-of-focus areas. While this greatly enhances resolution in the *z* direction (depth), and slightly in the *xy* direction, it also reduces signal strength significantly. Thus, confocal measurements commonly require higher laser power/longer exposure times than standard measurements.
66. There is a trade-off between signal-to-noise ratio improvements and feasible measurement times, especially if many points need to be recorded via raster scan mapping. Since signal-to-noise ratio can be easily and dramatically improved by multivariate noise filtering, it is generally preferred to keep exposure times to minimum, provided that at least a 100 points will be recorded in each image.
67. The step size by which to move in the *z* direction depends on the sample type, thickness, confocal settings and lens magnification. An example: when using a 20× magnification lens at normal confocality on 20 μm thick sections, usually 2 μm steps are a good starting value that can be refined iteratively. Smaller steps are required for higher magnifications and high confocality, as well as for refining.
68. Although the lasers themselves are already polarized, the entire setup needs to maintain and control the direction of the polarization, hence the need for polarizers.
69. Technically, a polarization stage is not a necessity, unlike the polarizer (*see* **Note 68**) but a convenience. While the sample can be rotated without a special stage, such rotations are harder to control and fine-tune their angle, ultimately leading to difficulties in obtaining (and reproducing) reliable results.
70. If there are two physical polarizers installed on the same mount (i.e., normal and half-plate, *see* **Note 38**), only one of them can be used at a time.
71. Changing the direction by, for example, rotating the sample (not recommended) or changing polarization direction (preferred) will influence the band intensities of those functional groups that are strictly ordered (e.g., crystalline cellulose chains of microfibrils). Functional groups that are disordered, i.e., have varying directions, will not show direction sensitive behavior, i.e., their band intensities will not vary.
72. Similarly to the lens setup (*see* **Note 44**), WiRE gets no information from the hardware as to which polarizer (if any) is applied and thus it has to be set manually.

73. Changing the polarization direction is easiest in the software. However, it is also possible by using the half-plate (if existing) or (theoretically) by rotating the sample by 90° on the polarization stage. However, sample rotation is the most difficult to achieve accurately, and is generally not recommended.

74. In case cosmic ray(s) would appear in the spectrum, click on the "Run" button again, until a spectrum free from cosmic rays is recorded. It is also possible to specify more than one scan during measurement setup and remove cosmic rays afterwards by comparing the different scans. Most of the times, however, it is quicker to simply rerun a single spectrum in case cosmic rays would appear in it (since no setup is required, just a simple click on the "Run" button, *see* **Note 64**), than wait several scans even if there are no cosmic rays in any of them.

75. While technically not necessary, it is good practice to give representative filenames, containing information both about the sample AND key measurement parameters (laser wavelength and power, lens, exposure time, laser power, confocality and polarization if applicable, or step size for image data). While these parameters are stored in the WiRE file structure, they can be tedious to find. However, if they are included in the filename, they immediately give a good overview of the parameters and can help setting up the next measurement by providing close to ideal starting values. An example filename could be: "Arabidopsis_hyp_514nm_10percent_1s_static1200_20x_5micronXY", denoting an image recorded over an Arabidopsis hypocotyl, using a 514 nm (green) laser at 10% power, 1 s exposure time, in static mode, centered at 1200 $cm^{-1}$, using 20× magnification and 5 μm step size in both the *X* and the *Y* direction. To avoid compatibility problems, do NOT use special characters in the filename, such as space, comma, and asterisk (Hence "percent" and not "%" in the filename example, and underscores instead of spaces or commas). If a file with the same name exists in the same location (folder), WiRE will ask whether it should be overwritten or not.

76. There is no point selecting a single point for a map. A line or an outline has limited use for area maps as well, leaving the options of filled rectangles or circles as the only real alternatives. While circular areas provide no real advantage over rectangular ones for normal xylem sample, dealing with their coordinates and layout in data processing steps is considerably harder. Thus, circular areas are to be avoided as well.

77. The right magnification lens must be selected in WiRE in order for the scaling to be correct (*see* **Note 44**).

78. The *X* and *Y* step sizes do not have to be equal or integers.

79. Stages with 100 nm accuracy are not uncommon, although such precision necessitates vibration-proof environments, including a vibration proof table.

80. Laser spot sizes depend on the lens and the laser frequency, as well as the general instrument and measurement setup. While oversampling (i.e., specifying smaller step sizes than the laser spot size) is possible and can be useful at times (e.g., super-resolution studies), it is unnecessary if such information is not utilized during data analysis.

81. When too many points cover the area, the gridline cannot be shown. This is a clear indication of unnecessary small step sizes or wrong magnification settings in the software (*see* **Note 44**). Please note also that points will be measured where the lines cross, and not in the center of the outlined small areas.

82. In raster scanning, measurement points will be recorded from left to right in the first row, then returning to the left of the second row and start again, proceeding to the right, and so on in each row thereafter. In snake mode scanning, however, the second row will start from the right, i.e., directly under the last point of the first row, and proceed to the left. When reaching the start of row two, it will proceed to row three and record it from left to right again. Thus, each row will be recorded in an alternating left-to-right right-to-left fashion. This makes pixel tracking outside of WiRE complicated, and only really has advantage if focus tracking is used, or if sudden, fast tray movements would introduce positional errors (the tray rapidly moving from the end of one row to the beginning of the next would result in sample gliding etc.). This is seldom the case for well prepared xylem sections.

83. While it is possible to manually set the *X* and *Y* starting and ending coordinates as well, not only the *X* and *Y* step sizes, in the dialog window, it is usually cumbersome to adjust the measurement area this way. Often, it is easier and faster to simply redraw the rectangle of the desired measurement area.

84. While *X* and *Y* step sizes can still be easily changed, it is NOT possible to draw a new rectangle at this stage, only to define a new area by its coordinates, which is usually not worthwhile.

85. Illumination can be freely changed now, including switching off any extra lights (*see* **Note 59**), since the white light image was already taken and will not be influenced any more.

86. During single spot measurements, many different settings are tested, most of which are destined to be discarded. Thus, it is not useful to automatically save those and clutter the hard drive with files that should only be deleted afterwards. For maps, however, those settings are already determined and thus the recorded data is likely to be kept.

87. While it can be helpful to select a band that represents an interesting feature (chemically or anatomically, i.e., lignin), it is NOT important to map any specific band at all. This feature will only be used to track the progress of the measurement and possibly to detect severe anomalies during measurement (out of focus areas, fluorescence problems, etc.), and NO analysis will be based on this integration. It is also possible to change the color of the generated integral map in the "Acquisition" tab. This is a purely cosmetic step with no practical consequences apart from aesthetics.
88. There could be several measurements defined and placed in the measurement queue. When all are set up, they can be run after each other automatically, instead of one by one as described in the present protocol. However, easy automation would require all mapped areas to be in the same sample AND at the same focal depth (*z* dimension), as WiRE may not be able to adjust for the *z* levels automatically (There are workaround ways to achieve this, including dummy depth profiling, but they are outside of the scope of the present protocol). It is, therefore, recommended to run each mapping measurement one by one. This way, each recorded image can be checked to be correct and containing useful information, before proceeding to the next area/sample (quality control).
89. The visualized integrals are scaled progressively, based on all intensities in all recorded spectra. Thus, fluorescence and cosmic rays may influence the scaling drastically.
90. The estimated times are approximate only and can be particularly error prone at the first couple spectra, with estimation accuracy improving as the measurement progresses.
91. If no filename has been specified during setup, the file must be saved using the "Save As…" command in the "File" menu, otherwise the recorded data will be permanently lost.
92. If no other windows are open, the "File" > "Open…" option can also be used. However, it is good practice to open the files in a new window.
93. WiRE uses the standard Windows dialog for file selection, allowing for multiple files to be selected at once, using the standard "Shift" or "Ctrl" + click markings for continuous or independent selections, respectively.
94. The same spectrum can be reached via the "View" tab as well, but it is more complicated and laborious, and thus not included in the present protocol for simplicity.
95. Sadly, only one spectrum can be exported at a time, making the process extremely tedious for large sets.

96. WiRE keeps the last used settings as default (for folder and file type) and will suggest the same filename as the original file had, with a warning message about overwriting data if the same file with the same extension already exists in the same location.
97. Since version 4, WiRE allows saving in Matlab format as well. This has considerable advantages for image files (reducing file size by removing redundancy) and spectral series (as their automated import to Matlab can be faster), but not for a single spectrum.
98. The resulting ".txt" files have standard ASCII format and can be opened in any text processing software, such as Notepad and Microsoft Word. They can also be directly imported (or copy pasted) into Microsoft Excel for plotting. This, however, is not recommended for image data files (*see* **Note 116**).
99. When a single spectrum is saved as ".txt" file, the first column of the file contains the wavenumbers and the second column contains the corresponding intensities.
100. Folder and file names should only consist of English alphanumeric characters (letters and numbers) and underscores. Keep in mind that Matlab is case sensitive, i.e., "Filename" does not equal "filename". It is recommended to keep files in their dedicated folder, not in the default Matlab folder, but avoid excessive subfolder nesting.
101. In fact, only the "Data" variable will be used in the data analysis steps, but since that cannot contain the file names, this variable is saved for keeping track of which spectrum is in which position in the "Data" variable.
102. It is critically important to remove all major cosmic rays before multivariate noise filtering is performed, since their presence would skew the principal component analysis.
103. Unfortunately, there is no way to directly go to a specific spectrum in the dataset, or to skip all markings from the current spectrum and proceed to the next spectrum. This means that if several cosmic rays are falsely marked, the user is forced to skip through ALL of them, one by one, before being allowed to proceed to the next spectrum. This is a major oversight in the software, since it can be extremely tedious to cycle through hundreds of noise spikes (*see* **Note 104**) to be able to filter off a single cosmic ray.
104. This can happen frequently, especially if neighboring pixels have different spectral features (baseline shape, intensity jump by, for example, fluorescence, or simply different chemical composition) and particularly in case of low signal-to-noise spectra.

105. It is usually a good option to chose an offset about half of the current value. There is a glitch in the software, however, that may produce an error message if the value is entered in a certain way. To avoid this, always adjust the values after the decimal dot first, and only change the integer parts afterwards if needed (e.g., to change from 1.3 to 0.7, change "3" to "7" first, then "1" to "0").
106. Some hot spots (deviating outlier) areas may still be visible. Thus, instead of a random noisy image, a few well-defined pixels remain.
107. Ideally, the variance explained drops drastically from one component to the next, clearly indicating the end point. However, this is not always the case, and there could be either a continuous, gradual decrease of the variation explained (instead of a sudden drop), or a drop followed by an increase (noticeable in the bottom left plot of the window as well, displaying the correlation of loadings as a function of the number of components).
108. For instance, if the end point was reached at component 5, the final number of component is 4 (since the fifth component only described noise and little variance). Thus "4" needs to be entered in the text box.
109. The value is not registered until the "Enter" key is pressed on the keyboard.
110. Unfortunately, the script introduces an artifact in the data: the uppermost wavenumber intensity will be erroneously low after reconstruction from PCA loadings. Thus, the spectral region will have to be trimmed before processing the spectra further, by cutting away ca. three data points (wavenumbers) at the high end of the spectra.
111. A backup of the original file is automatically generated during noise filtering, avoiding accidental overwriting. However, it is still recommended to specify a new file name, by for example adding "_nfX" to the end, indicating that this is the noise filtered version of the same file using X principal components.
112. Technically, several files can be opened at once, but since each needs to be processed individually, it is no real advantage of opening multiple files simultaneously, which can slow down less powerful computers.
113. Some options require spectral preprocessing (such as baseline correction), and are thus not available at this stage.
114. Maps can easily be removed from the list. Alternatively, they can be kept but made invisible by un-ticking their "Visible" checkbox, to facilitate visualizing.

115. There are further options as well, including the addition of a scale bar and showing axes. These are mostly straightforward and can be experimented with freely. They are available by the context sensitive menu that opens upon right-clicking on the map review image.

116. When an image file is saved as ".txt" file, WiRE is extremely wasteful and generates huge files with redundant information and suboptimal structures. The first column of the file contains the *Y* coordinates, the second contains the *X* coordinates, the third the wavenumbers and the fourth the corresponding intensities. Thus, all wavenumbers are repeated for each pixel, and the *X* and *Y* coordinates are repeated for each wavenumber in each spectrum.

117. Alternatively, the script can also be run from the EDITOR tab, by clicking on the "Run" button. In case the ".m" file is not in a folder included in the Matlab path, a dialog box opens, asking whether to change to that folder or to add that folder to the path. Select the change folder option.

118. In case of a series of spectra, there are no *X* and *Y* coordinates, as each spectrum is individual, not a pixel of an image. Thus, *X* and *Y* pixel numbers must be supplied manually. Dimensions can be chosen arbitrarily, but they need to match the total number of spectra in the dataset. For instance, if a dataset contains 30 spectra, $X=30$, $Y=1$ (the default suggestion by the script) is a valid choice, as are $X=5$, $Y=6$ and $X=3$, $Y=10$, etc. since $X \times Y=30$. It can be useful (particularly for large number of spectra) to use a divider that is meaningful from a data analysis point of view. For instance, if five spectra were taken from six biological replicates, $X=5$, $Y=6$ is a better choice than $X=3$, $Y=10$, since each row will represent one biological replicate.

119. If no files are selectable, change the open dialog box option to show all file types. On Macintosh computers, Mac OSX Finder may erroneously show selectable files as non-selectable (i.e., greyed out), even when "all files" has been selected for "type". However, it is only a display error and the script will be able to load the correct files.

120. To speed up the displaying step on slower computers, spectral plots contain maximum 100 spectra. If the original dataset contains more spectra, 100 of them are randomly selected for display.

121. If a white light image was saved (in for example **step 22** in Method 2 in Subheading 3.4.1) with the same filename as the image data file but ".jpg" extension, it is also loaded automatically for reference. If not, a white light image can be manually loaded at any stage.

122. The resulting ".mat" file does not contain information regarding *X* and *Y* coordinates anymore, even if an image ".txt" file was loaded originally. Thus, it is recommended to add this information to the filename upon saving, using the following format: "XXXxYYY_filename.mat". In this case, XXX and YYY denote the number of pixels in the *X* and *Y* dimensions, respectively. The Matlab script is constructed in a way that it will automatically use this information for *X* and *Y* dimensions and correctly folds the image data when next loaded. Since ".mat" files load considerably faster than the corresponding WiRE exported ".txt" files, it is highly recommended to save and use ".mat" files for image data over 200 pixels.

123. The original version of the script presented in reference [34] has built-in non-negativity constraints that cannot be changed. Moreover, it only allows the iteration to start by estimating spectral profiles (not concentrations), and does not allow for any other constraint to be applied. However, the newer version (available at http://www.kbc.umu.se/english/visp/download-visp/) is capable of starting the iteration from different starting profile estimates as well, and the analysis can be augmented by equality and unimodality constraints too. Nevertheless, for ultimate adaptability (including combined multimethod, multiimage datasets), a standalone MCR-ALS script, available at http://www.mcrals.info/ is recommended. The data saved after preprocessing from the current protocol can be directly used in that script without further modification. Spectra preprocessing, reference matching, clustering and visualisation, however, are not possible in that standalone MCR-ALS script.

124. Component maps should match anatomical/white light image features in image data sets, or correspond to sample differences in sets of spectra. Random maps or maps that do not reflect the biological features/sample differences can be products of suboptimal preprocessing or analysis. In such cases, a new model should be tested by changing preprocessing and/or key MCR-ALS parameters. Similarly, the resolved spectral profiles should be generally meaningful and never completely different from the features of the raw spectra.

125. Reference matching is based on Euclidean distances in the script.

126. In the latest version of the script, the main user interface window can be left open and a new dataset loaded via the dedicated button in the interface.

127. Matlab can use significant memory resources, thus it is recommended to clear all variables from the memory. Be aware, however, that all unsaved data will be lost after this stage, and the graphical user interface needs to be restarted for a new analysis.

## Acknowledgments

The protocols were developed and tested using the instruments of the Vibrational Spectroscopy Core Facility in Umeå, Sweden, with operational financial support from the Chemical Biological Centre (KBC) and the Department of Chemistry at Umeå University, Umeå, Sweden.

### References

1. Raman CV, Krishnan KS (1928) A new type of secondary radiation. Nature 121:501–502
2. Keresztury G (2001) Raman spectroscopy: theory. In: Chalmers JM, Griffiths PR (eds) Handbook of vibrational spectroscopy, vol 1: Theory and instrumentation, vol 1. John Wiley & Sons, Ltd, Chichester, England
3. Cheng J-X, Xie XS (2015) Vibrational spectroscopic imaging of living systems: an emerging platform for biology and medicine. Science 350(6264):aaa8870
4. Socrates G (2001) Infrared and Raman characteristic group frequencies. Tables and charts, 3rd edn. John Wiley & Sons, Ltd, Chichester, England
5. Gierlinger N, Schwanninger M (2007) The potential of Raman microscopy and Raman imaging in plant research—review. Spectroscopy 21:69–89
6. Gierlinger N, Keplinger T, Harrington M (2012) Imaging of plant cell walls by confocal Raman microscopy. Nat Protoc 7:1694–1708
7. Agarwal UP (1999) An overview of Raman spectroscopy as applied to lignocellulosic materials. In: Argyropoulos DS (ed) Advances in lignocellulosics characterization. TAPPI Press, Atlanta, GA, pp 201–225
8. Agarwal UP (2006) Raman imaging to investigate ultrastructure and composition of plant cell walls: distribution of lignin and cellulose in black spruce wood (*Picea mariana*). Planta 224:1141–1153
9. Agarwal UP, McSweeny JD, Ralph SA (2011) FT-Raman investigation of milled-wood lignins: softwood, hardwood, and chemically modified black spruce lignins. J Wood Chem Technol 31:324–344
10. Agarwal UP, Ralph SA (1997) FT-Raman spectroscopy of wood: identifying contributions of lignin and carbohydrate polymers in the spectrum of black spruce (*Picea mariana*). Appl Spectrosc 51:1648–1655
11. Chu L-Q, Masyuko R, Sweedler JV, Bohn PW (2010) Base-induced delignification of *Miscanthus*×*giganteus* studied by three-dimensional confocal Raman imaging. Bioresour Technol 101:4919–4925
12. Chylińska M, Szymańska-Chargot M, Zdunek A (2014) Imaging of polysaccharides in the tomato cell wall with Raman microspectroscopy. Plant Methods 10:14
13. Evans PA (1991) Differentiating "hard" from "soft" woods using Fourier transform infrared and Fourier transform Raman spectroscopy. Spectrochim Acta A 47:1441–1447
14. Hanninen T, Kontturi E, Vuorinen T (2011) Distribution of lignin and its coniferyl alcohol and coniferyl aldehyde groups in *Picea abies* and *Pinus sylvestris* as observed by Raman imaging. Phytochemistry 72:1889–1895
15. Himmelsbach DS, Khahili S, Akin DE (1999) Near-infrared–Fourier-transform–Raman microspectroscopic imaging of flax stems. Vib Spectrosc 19:361–367
16. Horvath L, Peszlen I, Gierlinger N, Peralta P, Kelley S, Csoka L (2012) Distribution of wood polymers within the cell wall of transgenic aspen imaged by Raman microscopy. Holzforschung 66:717–725
17. Lacayo CI, Malkin AJ, Holman HY, Chen L, Ding S-Y, Hwang MS, Thelen MP (2010) Imaging cell wall architecture in single *Zinnia elegans* tracheary elements. Plant Physiol 154:121–133
18. Larsen KL, Barsberg S (2010) Theoretical and Raman spectroscopic studies of phenolic lignin model monomers. J Phys Chem B 114: 8009–8021
19. Larsen KL, Barsberg S (2011) Environmental effects on the lignin model monomer, vanillyl alcohol, studied by Raman Spectroscopy. J Phys Chem B 115:11470–11480
20. Mathlouthi M, Koenig JL (1986) Vibrational spectra of carbohydrates. Adv Carbohydr Chem Biochem 44:7–89

21. Perera PN, Schmidt M, Chiang VL, Schuck PJ, Adams PD (2012) Raman-spectroscopy-based noninvasive microanalysis of native lignin structure. Anal Bioanal Chem 402: 983–987
22. Sun L, Varanasi P, Yang F, Loqué D, Simmons BA, Singh S (2012) Rapid determination of syringyl: guaiacyl ratios using FT-Raman spectroscopy. Biotechnol Bioeng 109: 647–656
23. Saariaho AM, Jääskeläinen AS, Matousek P, Towrie M, Parker AW, Vuorinen T (2004) Resonance Raman spectroscopy of highly fluorescing lignin containing chemical pulps: suppression of fluorescence with an optical Kerr gate. Holzforschung 58:82–90
24. Saariaho AM, Jääskeläinen AS, Nuopponen M, Vuorinen T (2003) Ultra violet resonance Raman spectroscopy in lignin analysis: determination of characteristic vibrations of p-hydroxyphenyl, guaiacyl, and syringyl lignin structures. Appl Spectrosc 57:58–66
25. Schulz H, Barańska M (2007) Identification and quantification of valuable plant substances by IR and Raman spectroscopy. Vib Spectrosc 1:13–25
26. Synytsya A, Copíková J, Matejka P, Machovic V (2003) Fourier transform Raman and infrared spectroscopy of pectins. Carbohydr Polym 54:97–106
27. Agarwal UP, Reiner RS, Ralph SA (2010) Cellulose I crystallinity determination using FT-Raman spectroscopy: univariate and multivariate methods. Cellul 17:721–733
28. Gierlinger N, Luss S, König C, Konnerth J, Eder M, Fratzl P (2010) Cellulose microfibril orientation of *Picea abies* and its variability at the micron-level determined by Raman imaging. J Exp Bot 61:587–595
29. Ackermann M et al (2013) Detection of the characteristic pion-decay signature in supernova remnants. Science 339:807–811
30. Diening T, Ibach W (2010) Software requirements and data analysis in confocal Raman microscopy. In: Diening T, Hollricher O, Toporski J (eds) Confocal Raman microscopy. Springer, New York
31. Savitzky A, Golay MJE (1964) Smoothing and differentiation of data by simplified least squares procedures. Anal Chem 36: 1627–1639
32. de Juan A, Maeder M, Hancewicz T, Duponchel L, Tauler R (2009) Chemometric tools for image analysis. In: Salzer R, Siesler HW (eds) Infrared and Raman spectroscopic imaging. Wiley-VCH, Weinheim, Germany
33. Grahn HF, Geladi P (eds) (2007) Techniques and applications of hyperspectral image analysis. John Wiley & Sons Ltd, Chichester, England
34. Felten J, Hall H, Jaumot J, Tauler R, de Juan A, Gorzsás A (2015) Vibrational spectroscopic image analysis of biological material using multivariate curve resolution–alternating least squares (MCR-ALS). Nat Protoc 10:217–240
35. Ramos PM, Ruisanchez I (2005) Noise and background removal in Raman spectra of ancient pigments using wavelet transform. J Raman Spectrosc 36:848–856
36. Jirasek A, Schulze G, Yu MML, Blades MW, Turner RFB (2004) Accuracy and precision of manual baseline determination. Appl Spectrosc 58:1488–1499
37. Eilers PHC (2004) Parametric time warping. Anal Chem 76:404–411
38. Baranska M, Schulz H, Rosch P, Strehle MA, Popp J (2004) Identification of secondary metabolites in medicinal and spice plants by NIR-FT-Raman microspectroscopic mapping. Analyst 129:926–930
39. Trygg J, Wold S (2002) Orthogonal projections to latent structures (O-PLS). J Chemometr 16:119–128
40. Gorzsás A, Stenlund H, Persson P, Trygg J, Sundberg B (2011) Cell-specific chemotyping and multivariate imaging by combined FT-IR microspectroscopy and orthogonal projections to latent structures (OPLS) analysis reveals the chemical landscape of secondary xylem. Plant J 66:903–914

# Chapter 13

# Using CellProfiler to Analyze and Quantify Vascular Morphology


## Liam Campbell, Manoj Kumar, and Simon Turner



## Abstract

Computational programs can be used in place of time-consuming, error-prone manual data collection. CellProfiler is a free, open source program that allows researchers to automate image analysis and collect large amounts of phenotypic data relatively easily. Here, we describe how to adapt CellProfiler to analyze cross sections of xylem tissue and use it to gather a variety of information on traits such as cell size, shape, and number. We provide step-by-step instructions to create a typical CellProfiler analysis pipeline, alongside explanations of important modules, options and parameters available to the user.



**Key words** CellProfiler, Image analysis, Bioinformatics, Xylem morphology, Lumen, Cell wall


## 1 Introduction

Xylem-related phenotypes can present in a variety of ways. Cell organization, size, number, shape, and identity can all be disrupted and frequently more than one of these traits are affected simultaneously. However, quantifying these changes can be problematic, usually requiring repetitive and time-consuming manual data collection that is susceptible to sampling bias and error. Fortunately, there are now a number of programs that can take advantage of modern computational power to reliably gather large amounts of data from biological images, yet recent publication suggest these computational methods are applied relatively rarely in studies of vascular development.

CellProfiler [1] is an intuitive and easy to use image analysis program requiring little prior bioinformatics knowledge. The program is free, open source, frequently updated and contains a straightforward modular pipeline-based user interface. Though other programs such as ImageJ [2] and Bitplane Imaris may have more advanced features such as 3D tracking, xylem image analysis does not usually require such complexities and the relative simplicity of CellProfiler facilitates easy analysis. By November 2015, the


Miguel de Lucas and J. Peter Etchells (eds.), *Xylem: Methods and Protocols*, Methods in Molecular Biology, vol. 1544,
DOI 10.1007/978-1-4939-6722-3_13,

program had been cited 2859 times in the literature and has been adapted for a wide array of purposes ranging from worm tracking [3] to eye pigment analysis [4] and automated seed counting [5].

In this chapter, we will introduce the reader to the basics of CellProfiler through an in-depth example that will demonstrate how to create a pipeline to analyze xylem cross sections and output measurements such as cell size, shape and number. We will also provide other examples of what can be achieved with the program, including the quantification of a mutant phenotype, but it must be stressed that these are only suggestions and they should not be carelessly assumed to work for every individual use. The reader will have to modify or create their own pipelines that are specifically designed to work for the images at hand. Similarly, the examples we show have been developed to measure particular traits that are of interest to us, but clearly CellProfiler can be adapted to measure a much wider variety of traits. We hope that the examples presented give an insight into the general image analysis procedure and encourage other researchers to adapt CellProfiler to measure traits that maybe of particular interest to them.

## 2 Materials

The reader should download and install the latest version of CellProfiler (2.2.0 at the time of publication) from the CellProfiler website (www.cellprofiler.org). CellProfiler is available for Windows, Mac and Linux. Images should be of good quality, stained histological sections with few tears or other forms of damage are ideal (*see* **Notes 1** and **2**). The magnification used should be identical for each image and be of a high enough power to be able to resolve individual cells by eye. The image format should also be consistent as this can alter the output accuracy of the program. We recommend the TIFF format as this retains full detail due to an absence of image compression, unlike some more common formats such as JPEG.

## 3 Methods

### 3.1 CellProfiler—A General Overview

Upon start-up, by default the user is presented with an empty pipeline. Figure 1 demonstrates the basic CellProfiler navigational elements. The flowchart in Fig. 2 shows a typical CellProfiler workflow, demonstrating key steps and modules likely to be found in most xylem analysis pipelines. To summarize, inputted images are processed to images more suitable for object (cell) identification, such as a binary (black and white) image. CellProfiler then identifies objects within the processed images, measures these and outputs the data to a Microsoft Excel workbook for simple analysis.

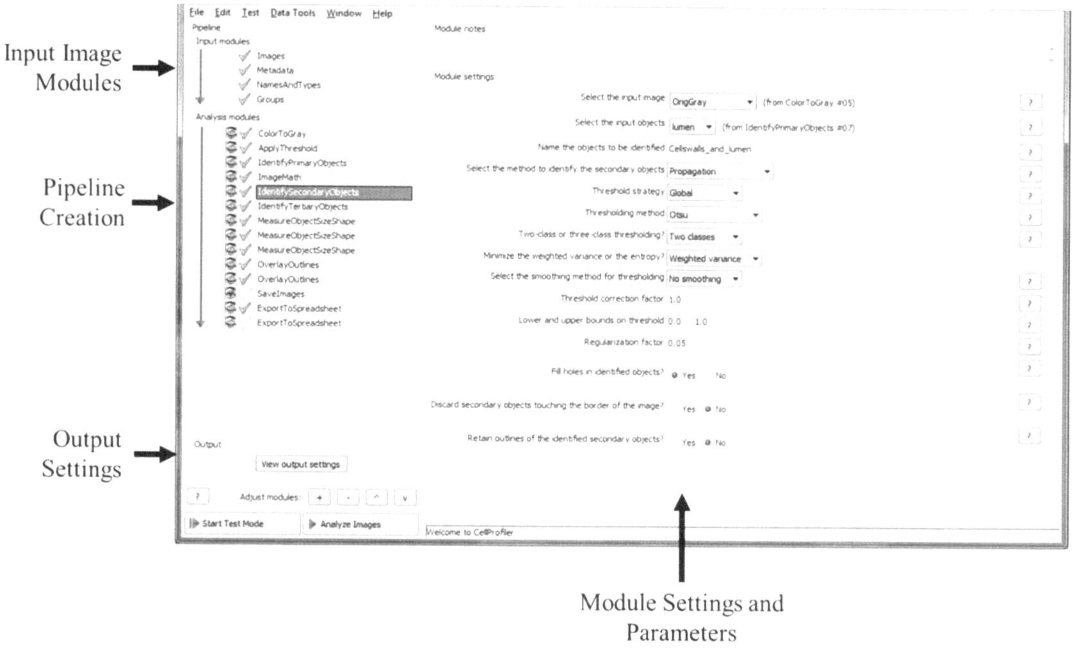


**Fig. 1** Overview of CellProfiler. The top left of the program is reserved for input settings, where the images to be analyzed are added and categorized if necessary. Directly below is the pipeline creation box. Here, modules are inserted and ordered in a manner that will take the inputted images and process them step-by-step as the user requires. When a module is selected, detailed settings for that module are shown in the large space on the right of the screen. This is where a variety of parameters can be modified to suit the pipeline being created. Choosing the optimum parameters for any given pipeline often requires trial-and-error. Below the pipeline at the bottom left of the program are the output settings in which the user identifies where on their computer the final measurements/processed images are to be saved. Also located here are the 'Adjust Modules' icons, used to insert/delete/order individual modules

Though these major steps are likely to be found in most pipelines, there are many additional modules that can be incorporated throughout.

### 3.2 Cropping Images

Before images are analyzed in CellProfiler, they should be cropped to leave only the region of interest. In the case of the example pipeline below (Subheading 3.3), this is a cropped region of secondary xylem tissue from a toluidine-blue stained Poplar stem section (Fig. 3a). Larger regions will obviously produce more accurate data but take longer to analyze. Almost any common image-editing program will be able to crop images. ImageJ and GIMP are both freely available and user-friendly.

### 3.3 Example Pipeline—Identifying and Measuring Xylem Cells and Cell Walls

The procedure below will take the reader through the creation of a basic pipeline that allows for the identification and measurement of the cells and cell walls contained within a typical secondary xylem section. As we cannot provide a full explanation of every term, module or setting used, please refer to the documentation

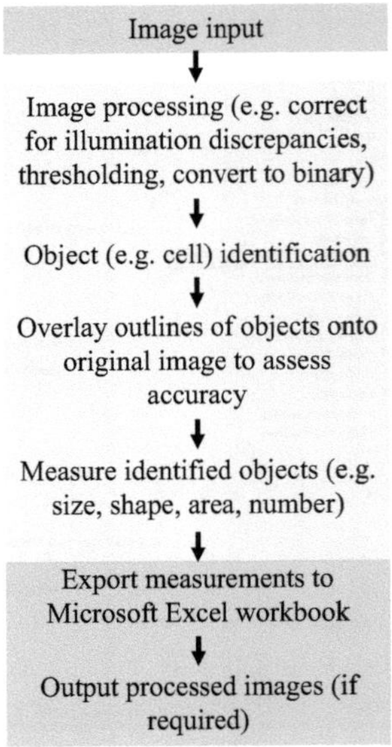


**Fig. 2** A typical cell analysis pipeline. Input images are preprocessed and cells are identified and measured. The measurements are exported to a spreadsheet to allow for downstream data analysis

and tutorials on the CellProfiler website (www.cellprofiler.org) for more comprehensive information if required. Figure 3a provides an overview of the pipeline described and should be used as a visual reference for each of the steps described below.

1. Insert a few sample color section images (*see* **Note 3**) into the '*Images*' drag-and-drop box within the input modules at the top left of the program window. In the '*NamesAndTypes*' input module, rename these input images to 'Colour'. This will help you to track the original color images at later stages.
2. Click the 'Add Module' button at the bottom left of the program window and insert a '*ColorToGray*' module, found within the 'Image Processing' tab. Set the conversion method to combine, the output image name to 'Grayscale' and select the original color images as the input (the images added in **step 1**). This will create a grayscale copy of your images for further processing.
3. Add an '*ApplyThreshold*' module, again from 'Image Processing'. Ensure that the input is set to the grayscale image from **step 2**. Select 'Binary' as the output image type and rename the output accordingly. This will create a binary (black and white) copy of your grayscale images that is more suitable for cell identification. If the binary image does not closely resemble the actual image, for example if some cells have been lost, you may want to alter the threshold strategy and parameters until you find a suitable setting for your images. The documentation provided by CellProfiler will aid this (*see* **Note 4**).

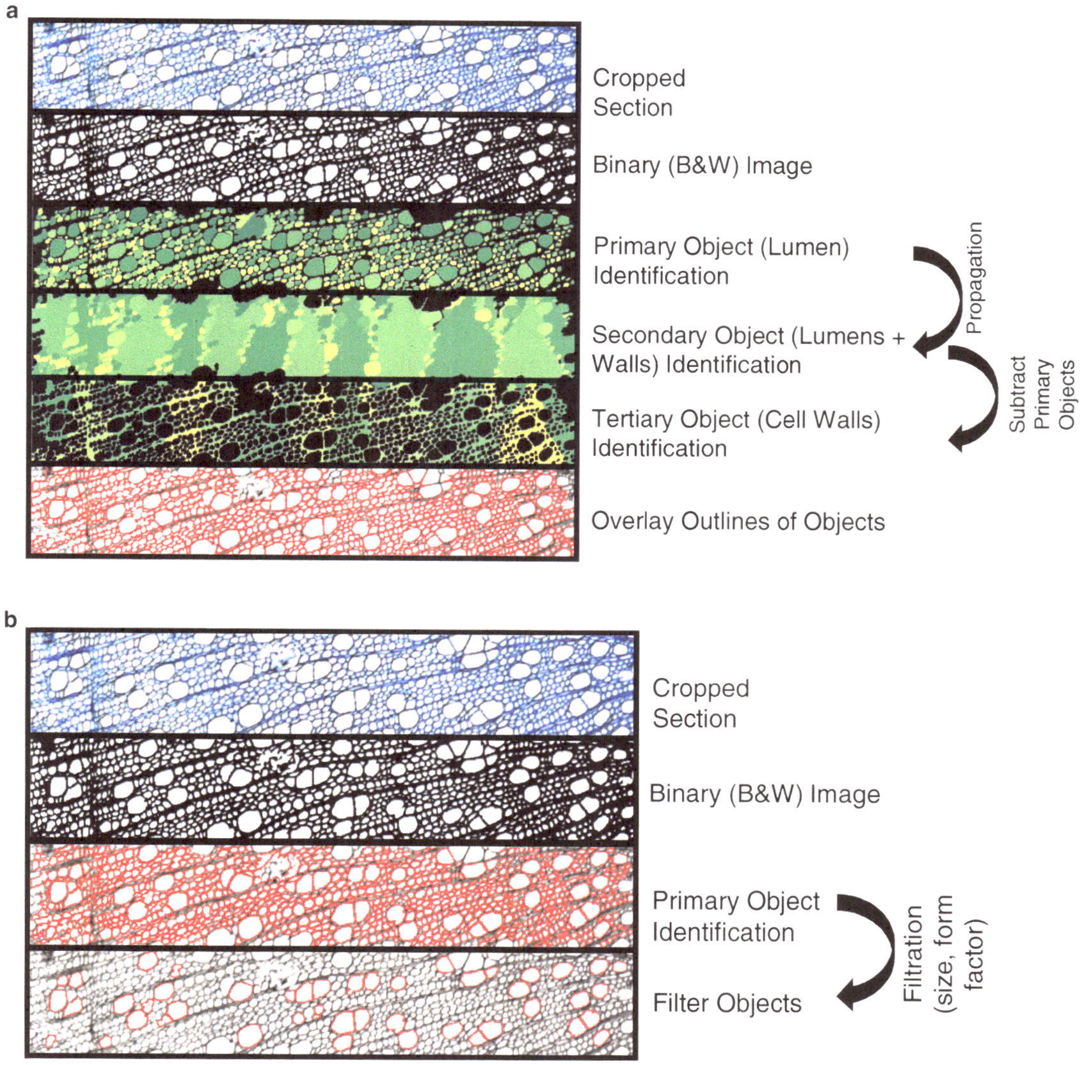


**Fig. 3** Visual representations of pipelines to identify and measure xylem lumens and walls from a Poplar hypocotyl. (**a**) The section is cropped and converted to binary. Primary objects (cell lumens) are identified and propagated outwards to form secondary objects which also include the cell walls. The primary objects are removed from these secondary objects to leave only the walls. The lumen and wall properties can then be measured separately. An overlay of the identified cells can be placed on top of the original image to assess identification accuracy. (**b**) Following primary object/cell identification, a filtration step can be added to separate cell types based on size and shape. In this case, vessels are automatically distinguished from fibers. Figures reproduced and adapted from [7] with permission from Elsevier

4. From the 'Object Processing' tab within 'Add Modules', insert the '*IdentifyPrimaryObjects*' module. This will be used to identify cell lumens within your images and is a critical step upon which the rest of the pipeline will rely (*see* **Note 5**). The input should be the binary image from **step 3** and rename the primary objects to 'Lumens'. To deduce what the typical diameter for lumens in your images are, run the pipeline created so far.

Within any image output window, click '*Tools*' → '*Measure Length*' and drag over some sample cells to gauge their diameter. Return to the main program window and enter provisional *min* and *max* values for lumen size. Automatic thresholding works well for most images and all other settings can be left as default. You will want to retain the outlines of the identified objects for use in the next step, so tick this box towards the bottom.

5. Add an '*OverlayOutlines*' module from within 'Image Processing' and set it to place the outlines of cells identified in **step 4** on top of the grayscale or color image. The images created by this module will reveal how accurately the pipeline has identified cells. Run the pipeline and assess accuracy. If cells are clearly being excluded or misidentified, adjust the *min* and *max* boundaries accordingly and repeat until suitable values are found. This may take some trial-and-error (*see* **Notes 6** and 7).
6. Add an '*IdentifySecondaryObjects*' module. Set the method of object identification to propagation from the primary objects. Automatic or global Otsu thresholding usually gives good results for xylem tissue. Rename the identified objects in this module to 'Cell_Walls_&_Lumens'. This module will propagate outwards from the identified lumens to create new objects that covers both the cell lumens and walls. Run the pipeline and evaluate whether both cell walls and lumens are being identified. If not, refer to the documentation and attempt other thresholding parameters accordingly.
7. To subtract the lumens (**steps 5** and **6**) from the cell walls and lumens from **step** 7, leaving only the cell walls, add in an '*IdentifyTertiaryObjects*' module. For the 'larger objects' select the 'Cell_Walls_&_Lumens' from **step** 7, and for the 'smaller objects' select the 'Lumens' identified in **steps 5** and **6**. Rename the identified objects to 'Cell_Walls_Only'. Run and evaluate the pipeline. If the previous modules have been correctly set up there should be few issues with this step.
8. At this point you should have a pipeline that has converted your inputted images to binary form and accurately identified the cell lumens and walls within the image. If not, the most likely sources of error are the thresholding to binary and the primary object (lumen) identification parameters.
9. Insert two '*MeasureObjectSizeShape*' modules from within the 'Measurement' tab (*see* **Note 8**). Assign one to measure the primary objects, 'Lumens', and the second to measure the tertiary objects, 'Cell_Walls_Only'. The former will measure a variety of characteristics for each individual cell in the image. This includes but is not limited to the total number of cells and the area of each cell (*see* **Note 9**). The latter will measure the

total area of cell wall material in the image. Because of a lack of boundary between adjacent cell walls, the program cannot distinguish walls individually. However, a simple calculation of '*Total Cell Wall Area/Total Number of Cells*' will return the average cell wall area per cell.

10. Insert an '*ExportToSpreadsheet*' module which will export the measurements to a spreadsheet within the specified output folder when the pipeline is run. Within the module settings you can define what measurements to export, if all are not needed, as well as requesting that the program creates automatic mean, median, and standard deviation values for each measurement. A single '*ExportToSpreadsheet*' module can export measurements for all identified modules.

### 3.4 Example Pipeline—Filtering Objects to Identify Xylem Vessels

In some cases, only certain cell types are of interest to a researcher. For example, a candidate gene may be suspected to regulate xylary vessel development without affecting other cell types. In such situations, CellProfiler can be adapted to separate and filter out certain cell types from a cross section and return measurements from only these. The steps below will guide the reader through a simple pipeline that filters only vessel cells from a secondary xylem cross section whilst ignoring fiber and parenchyma cells. Figure 3b provides a visual representation of this pipeline.

1. Follow **steps 1–6** from the pipeline described in Example 3.3, to the point at which all cells in the image are being identified with good accuracy (*see* **Note 10**).
2. The major distinguishing property between xylem vessels and fibers is cell area, so during the early pipeline creation phase, use '*Tools→ Measure Length*' within any open image window to get a gauge of the typical vessel diameter in your sections. This will be used in **step 4**.
3. Insert a '*MeasureObjectSizeShape*' module to measure the identified lumens.
4. Insert a '*FilterObjects*' module from the 'Object Processing' page. Rename the identified objects to 'Vessels' and select to filter lumens based on measurements.
5. Choose to filter by primary object area and based upon the diameters of the vessels measured in **step 2**, enter preliminary *min* and *max* area values for vessel filtration.
6. Towards the bottom of the '*FilterObjects*' settings, choose to save the outlines of the filtered objects, rename these to 'VesselOutlines' and insert an '*OverlayOutlines*' module to overlay these on top of your original image.
7. Run the pipeline and assess whether vessels are being accurately identified (*see* **Notes 11** and **12**). If not, alter the *min* and *max* filter boundaries accordingly and repeat until suitable values are found.

### 3.5 Example Application of CellProfiler—Quantification of the 'irregular xylem' (irx) Phenotype

The collapsed xylem or '*irx*' phenotype was first described by Turner and Somerville in 1997 [6]. Since then, mutations in a number of genes involved in biosynthesis of secondary cell wall components have been shown to cause this phenotype. These include genes involved in cellulose (*IRX1*, *2*, *3*, *5*, and *6*), xylan (*IRX7*, *8*, *9*, *10*, *14*, and *15*) and lignin (*IRX4*, *12*) biosynthesis as well as regulatory genes (*IRX11*). There can be significant variation among the *irx* alleles and it is therefore useful to computationally quantify their phenotype. We have used CellProfiler to calculate a measure of *irx* phenotype severity, as described below.

1. Crop vascular bundles from toluidine blue stained images of Arabidopsis stems.
2. (Optional). As a test case, compare wild type Ler-0 and the strong mutant *cesa7/irx3-1*. Of the many measurements outputted by CellProfiler, it was found that 'form factor' could be used as an indicator of cell deformity. Form factor is calculated as $4 \times \pi \times \text{area}/\text{perimeter}^2$ and is a general measure of roundness. A lower form factor in this case indicates a more deformed cell, or a more severe *irx* phenotype.
3. Plot the distribution of form factor measurements (those from Ler-0 and *irx3-1* are shown in Fig. 4a).
4. (Optional). The smallest cells (Fig. 4b) can be problematic to analyze. Filter out if required.
5. View form-factor distributions to classify the *irx* phenotype. We use a form factor of <0.7 to classify tissue as having an *irx* phenotype. An example bundle is shown in Fig. 4c, demonstrating cells which would pass the threshold and those that would not. An 'IRX Score', i.e., the percentage of cells with a form factor of <0.7 (Fig. 4d), can be used to aid interpretation, e.g., Ler-0 wild type has only about 15 % cells with a form factor of <0.7 (IRX Score 15), whereas in *cesa7irx3-1* the proportion of such cells is 73 % (IRX Score 73).

## 4 Notes

1. The type of stain and length of staining used on the section can affect the results obtained by CellProfiler. Higher contrast stains are more likely to produce accurate cell identification by making the boundary between cells more distinct. If you are struggling to distinguish cells by eye, the program will likely struggle to distinguish them also. Experiment with different stains to see which works best for you. We used toluidine blue for all images presented in this chapter. Microtome-created sections from fixed tissue are ideal but hand sections from fresh material can also be analyzed.

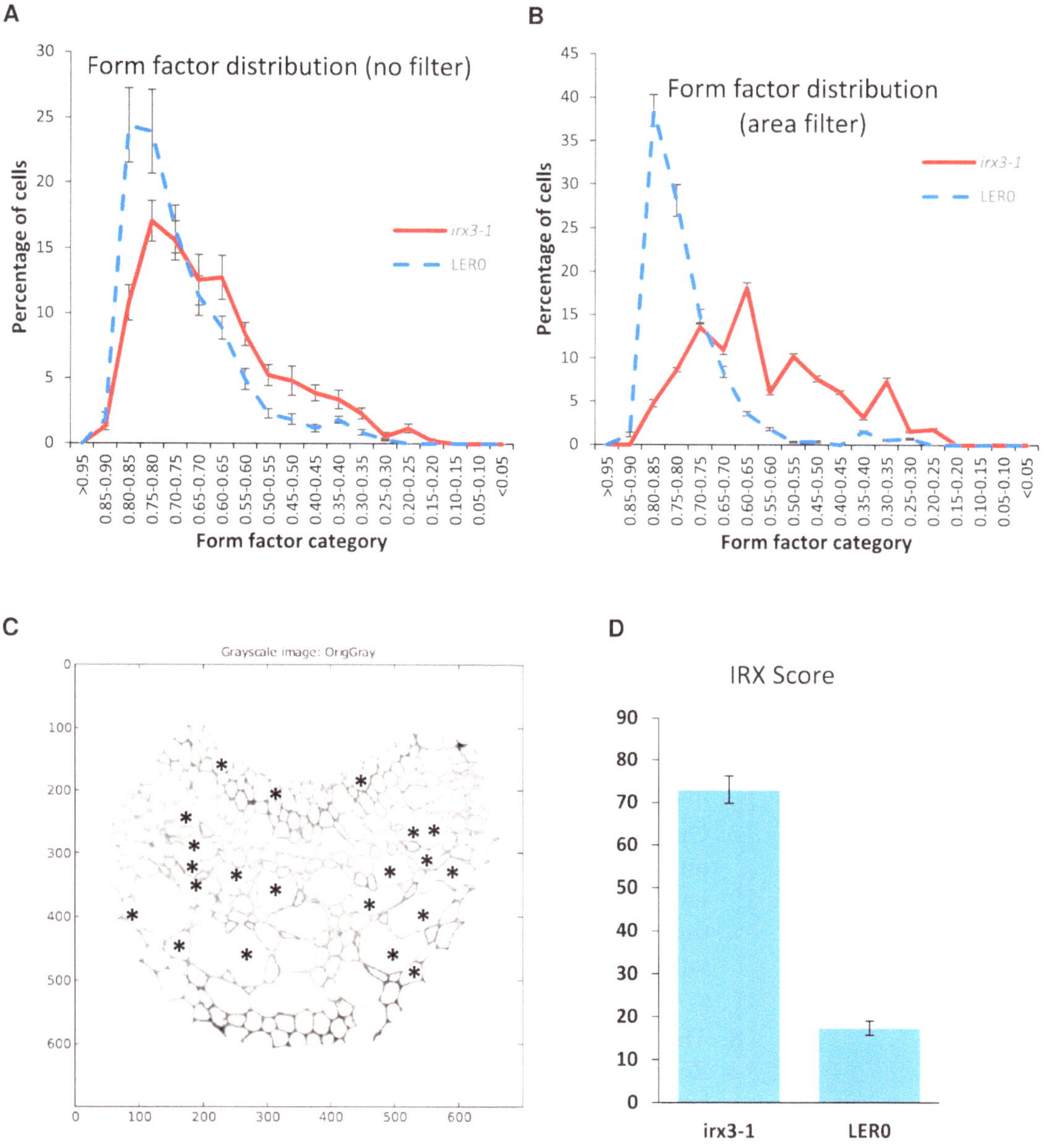


**Fig. 4** Using CellProfiler for quantification of *irx* phenotype. Vascular bundles were isolated from three Ler0 plants (13 bundles in total) and two *cesa7irx3-1* plants (seven bundles in total). Form factor was chosen as an indicator of cell deformity. The distribution of form factor in the two genotypes analyzed was plotted using analysis of all cells (**a**) or after filtering out the smallest cells (**b**). Clear differences can be seen between two genotypes at form factor of 0.7. An example bundle with a mixture of cells (**c**) is shown with *asterisks* marking cells with form factor <0.7. IRX score (**d**) was defined as percentage of cells with form factor <0.7

2. There are a variety of image illumination and artefact correction tools available within CellProfiler that can help to remove issues related to image quality such as uneven illumination.

3. We recommend using a small number of representative images (2–3) during pipeline creation to reduce the time it takes to construct a pipeline tailored to your image set. Of course,

before trusting any output measurements you should check that your pipeline is performing consistently across all of your images at each step (use the eyeball icon next to each).

4. There are detailed explanations of all modules and module parameters available on the CellProfiler website alongside video tutorials, example pipelines and a discussion forum. Additionally, clicking the '?' to the right of each parameter will provide you with a good summary of what the parameter in question is.
5. The default '*IdentifyPrimaryObjects*' module is set to discard objects/cells crossing the image boundary. This ensures that measurements such as cell size are not taken for what may only be a small portion of the actual cell. You may want to remove this setting in a minority of cases but we recommend leaving this as default to reduce potential measurement errors. Though data for some cells are lost, you can be more confident in the measurements produced.
6. It is always a good idea to check that the outputs from your pipeline make biological sense. For example, area measurements for vessel cells will tend to be larger than for fibers. Measurements can be easily overlaid on top of individual objects/cells by using the '*DisplayDataOnImage*' module. This is a simple and effective way of evaluating output measurements.
7. Once satisfied that cells are being accurately identified you can untick the eye-ball next to all modules added so far. This will save time, computational power and clutter by not bringing up the visual output for these modules, for which you are already satisfied with the results and therefore need no more evaluation.
8. The '*MeasureObjectSizeShape*' module is set to calculate Zernike features by default. These are usually not required for basic xylem analysis and can be turned off to save a substantial amount of analysis time.
9. All measurements returned by CellProfiler are calculated in pixel units by default. For example, cell area measurements are measured as $pixels^2$. You may convert these to a more relevant unit if required.
10. You can use the 'Test Mode', located at the bottom left of the program window, to quicken pipeline creation by allowing you to proceed through a small number of modules at a time, rather than the whole pipeline having to be run at once.
11. As an additional safeguard to prevent artefacts such as tears from being misidentified as vessels, you may want to add an additional filter within the '*FilterObjects*' module to filter based

on cell shape. Vessels are generally quite circular and so 'form factor' boundaries of 0.6-1 are suggested.

12. The vessel filtration is unlikely to be 100% accurate due to the variation of size within cell types—some larger fibers will be misidentified vessels and vice versa. We would suggest that minimizing false positives should take precedence over reducing false negatives as the remaining identified cells from measurements would be taken from are very likely to be vessels, even if some smaller vessels are missed.

## References


1. Carpenter AE, Jones TR, Lamprecht MR, Clarke C, Kang IH, Friman O, Guertin DA, Chang JH, Lindquist RA, Moffat J, Golland P, Sabatini DM (2006) Cell Profiler: image analysis software for identifying and quantifying cell phenotypes. Genome Biol 7(10):100
2. Rasband WS (1997) ImageJ. National Institutes of Health, Bethesda, MD, USA. Available from: http://imagej.nih.gov/ij/
3. Wählby C, Kamentsky L, Liu ZH, Riklin-Raviv T, Conery AL, O'Rourke EJ, Sokolnicki KL, Visvikis O, Ljosa V, Irazoqui JE, Golland P, Ruvkun G, Ausubel FM, Carpenter AE (2012) An image analysis toolbox for high-throughput *C. elegans* assays. Nat Methods 9(7):714–716
4. Jiang Y, Qi X, Chrenek MA, Gardner C, Boatright JH, Grossniklaus HE, Nickerson JM (2013) Functional principal component analysis reveals discriminating categories of retinal pigment epithelial morphology in mice. Invest Ophthalmol Vis Sci 54(12):7274–7283
5. Mussadiq Z, Laszlo B, Helyes L, Gyuricza C (2015) Evaluation and comparison of open source program solutions for automatic seed counting on digital images. Comput Electron Agric 117:194–199
6. Turner SR, Somerville CR (1997) Collapsed xylem phenotype of Arabidopsis identifies mutants deficient in cellulose deposition in the secondary cell wall. Plant Cell 9(5):689–701
7. Etchells JP, Mishra LS, Kumar M, Campbell L, Turner SR (2015) Wood formation in trees is increased by manipulating PXY-regulated cell division. Curr Biol 25:1050–1055

# Part III

## Composition

# Chapter 14

# Lignin Analysis by HPLC and FTIR

Jorge Reyes-Rivera and Teresa Terrazas

## Abstract

Fourier transform infrared spectroscopy (FTIR) is a simple non-destructive technique which allows the user to obtain quick and accurate information about the structure of the constituents of wood. High performance liquid chromatography (HPLC) is an analytical technique useful to determine the ratio of the lignin monomers obtained by the alkaline nitrobenzene oxidation method. Furthermore, lignin content has been commonly determined by wet chemical methods; Klason lignin determination is a quick and accessible method. Here, we detail the procedures for chemical analysis of the wood lignin using these techniques.

**Key words** FTIR, HPLC, Klason lignin determination, Lignin, Wood

## 1 Introduction

The chemical composition of secondary xylem (wood) has been extensively studied in economically important species [1–3]. Because lignin is the second most abundant component in wood after cellulose, it has been widely studied to determine its influence on the structural properties of wood [4–7]. A considerable number of techniques have been developed and applied to the qualitative and quantitative analysis of lignin [7]. Among the most frequently used analytical methods are Fourier transform infrared spectroscopy (FTIR) [8–26] and high performance liquid chromatography (HPLC) [20, 22, 27, 28]. Table 1 briefly summarizes their applications. However, lignin content has also been commonly determined by wet chemical methods, with Klason lignin analysis being the most frequently used [21, 22, 28–33].

FTIR spectroscopy has been used as a simple technique for obtaining quick and accurate information about the structure of the constituents of wood and the chemical changes that these constituents undergo during developmental or degradation processes [7, 22, 44, 46–48]. This non-destructive technique has multiple advantages, such as high sensitivity and selectivity, high signal/

Miguel de Lucas and J. Peter Etchells (eds.), *Xylem: Methods and Protocols*, Methods in Molecular Biology, vol. 1544,
DOI 10.1007/978-1-4939-6722-3_14, 

**Table 1**
**Applications and complements for lignin analysis by FTIR and HPLC**

| Method/Analysis | Applications | References |
|---|---|---|
| FTIR | Lignin structure<br>Lignin monomers | [8–22]<br>[14, 19, 21–26] |
| HPLC | Lignin structure<br>Lignin monomers | [27]<br>[22, 28] |
| Klason lignin | Lignin content | [21, 22, 27, 29–33] |
| Nitrobenzene oxidation | Lignin monomers | [5, 13, 22, 28, 34–37] |
| *Evolutive inference* | | |
| FTIR, HPLC, anatomy, and Klason lignin | Comparative, qualitative and quantitative: lignin content, lignin monomers, structure–function relationships | [22] |
| *Statistical complements* | | |
| FTIR, RL | Quantitative: lignin content | [38] |
| FTIR, MLR | Quantitative: lignin content, celluloses and hemicellulose | [39] |
| FTIR, PLS | Quantitative: percentage of acetyl groups | [40] |
| | Quantitative: lignin content, celluloses, hemicellulose and extractives | [41] |
| | Quantitative: lignin content (Kraft method) | [42] |
| FTIR, PLS, PCR | Quantitative: hydroxyl groups content | [43] |
| | Quantitative: lignin content prediction | [44] |
| FTIR, PCA | Qualitative: Characterization of tropical woods | [10] |
| | Qualitative: lignin characterization | [19] |
| | Qualitative: separation of species | [45] |
| FTIR, PCA, HCA, PLS | Qualitative and quantitative: lignin content prediction and characterization of woods | [46] |

noise ratio, high accuracy, ease of data handling, simplicity of the equipment required, short analysis times (minutes, compared with several days, as required by other methods), and the use of small sample sizes (a few milligrams, compared with several grams as required by conventional gravimetric techniques) [7, 13, 46]. In addition, the FTIR spectrum of a lignin sample provides an overview of its chemical structure [6, 9, 21, 22]. HPLC is also a non-destructive analytical method that allows for the separation and analysis of non-volatile or thermally unstable molecules with various polarities, as well as the recovery of the fractions in the sample [46]. Thus, this analytical technique has been useful to determine the ratio of the lignin monomers (syringyl/guaiacyl (S/G)) obtained by the alkaline nitrobenzene oxidation method [5, 13, 22, 28, 34–37].

The range of applications of FTIR spectroscopy is very broad. This technique is useful in the study of changes in the chemical composition of wood, as shown by the peaks (assigned to lignins or carbohydrates as described in the literature, Table 2) obtained in the FTIR spectrum of a sample [9, 20, 22, 41, 47, 48]. In addition, the application of chemometrics to the data resulting from FTIR spectroscopic studies is a powerful tool for both qualitative and quantitative analyses. The use of linear regression and multiple linear regression analyses allows for the determination of the content of lignin, cellulose and hemicelluloses in different lignocellulosic materials, including wood and pulp [38, 39]. Furthermore, partial least squares regression (PLS) and principal components regression (PCR) can be used to help predict lignin content [40–44, 46]. In addition, qualitative analyses using PLS combined with principal component analysis (PCA) and hierarchical cluster analysis (HCA) can be used to characterize different types of wood and lignins [10, 19, 45, 46].

Recently, the structural analysis of lignin by FTIR spectroscopy and the analysis of the S/G ratio by HPLC allowed us to relate the lignification of wood with morpho-anatomical and evolutionary aspects of species [22]. We found differences in the lignification of species with contrasting growth forms (Fig. 1a and b) due to the anatomical variations of the wood of different species and the support the lignified cells give the stem (Fig. 2). Here, we detail the procedures for determining the Klason lignin content of samples and their subsequent analysis by FTIR spectroscopy, as well as those for the isolation of lignin subunits by the nitrobenzene oxidation method and their analysis by HPLC.

## 2 Materials

Prepare all solutions using ultrapure water and analytical grade reagents.

### *2.1 Sampling and Extractive-Free Wood Isolation*

1. Mini mill, e.g., (catalog number 3383-L60; Thomas Wiley).
2. Ethanol–benzene mixture (1:2): mix together 1 volume of ethanol and 2 volumes of benzene (*see* **Note 1**).
3. Ethanol: 96% solution in distilled water.
4. Soxhlet apparatus: a heating mantle (HM), a round bottom flask (RBF), a Soxhlet extractor (SE), extraction thimbles (ET), an Allihn condenser (AC), and a cold circulator (CC). For assembly *see* Fig. 3a).

**Table 2**
**Assignment of bands in the FTIR spectra of wood samples**

| Wave no. ($cm^{-1}$) | Assignments | References |
|---|---|---|
| 3400–3430 | O–H stretching | [7, 9, 49, 50] |
| 2930 | $CH_2$ asymmetric vibration (guaiacyl-syringyl) | [47, 48] |
| 2850 | C–H stretching in methyl and methylene groups | [49] |
| 1734–1738 | Unconjugated C=O in xylans (hemicellulose) | [7, 9, 49, 50] |
| 1715 | Carbonyl stretching in conjugated ketone and conjugated carboxylic groups | [49, 50] |
| 1650 | Absorbed O–H and conjugated C–O | [9] |
| 1610 | C=C aromatic ring vibration | [49] |
| 1501–1511 | C=C aromatic ring vibration (guaiacyl-syringyl) | [7, 9, 45–48, 50] |
| 1463 | $CH_2$ deformation stretching in lignin and xylan | [46] |
| 1460 | C–H asymmetric deformation | [13, 50] |
| 1425 | Aromatic skeletal combined with C–H in-plane deforming and stretching | [46, 49] |
| 1375 | C–H deformation in cellulose and hemicellulose, and aliphatic C–H stretching in methyl and phenol OH | [9, 46] |
| 1330 | C–H vibration in cellulose | [9] |
| 1325 | Syringil ring breathing with C-O stretching | [13, 51] |
| 1317–1321 | Condensation of guaiacyl and syringyl units, syringyl unit and $CH_2$ bending stretching (softwood and hardwood, respectively) | [46] |
| 1285 | C–O and glucopyranosic cycle guaiacylic symmetric vibration | [47] |
| 1268 | Guaiacyl ring breathing, C–O stretching in lignin and C–O linkage in guiacyl aromatic methoxyl groups | [9, 45, 46, 51] |
| 1244 | Syringyl ring and C–O stretching in lignin and xylan | [9] |

| | | |
|---|---|---|
| 1225 | C–O and glucopyranosic cycle syringilic symmetric vibration | [47, 49, 50] |
| 1160 | C–O–C stretching in pyranose rings, C–O stretching in aliphatic groups | [46] |
| 1158 | C–O–C vibration in cellulose and hemicellulose | [9] |
| 1151 | CH stretching in aromatic ring (guaiacylic) | [7] |
| 1122 | Aromatic skeletal and C–O stretching | [9] |
| 1115–1116 | C–O–C stretching and symmetric vibration of the ester linkage, and CH stretching in aromatic ring (syringylic) | [7, 52] |
| 1085 | C–O deformation in secondary alcohol and aliphatic ether | [49] |
| 1048 | C–O stretch in cellulose and hemicellulose | [9] |
| 1030 | C–H in-plane deformation in guaiacyl and C–O deformation in primary alcohol | [46, 49, 53] |
| 913 | =CH out-of-plane deformation in aromatic ring (guaiacylic-syringylic) | [48] |
| 898 | C–H deformation in cellulose | [9, 46] |
| 855 | Aromatic C–H out-of-plane deformation | [13, 49] |

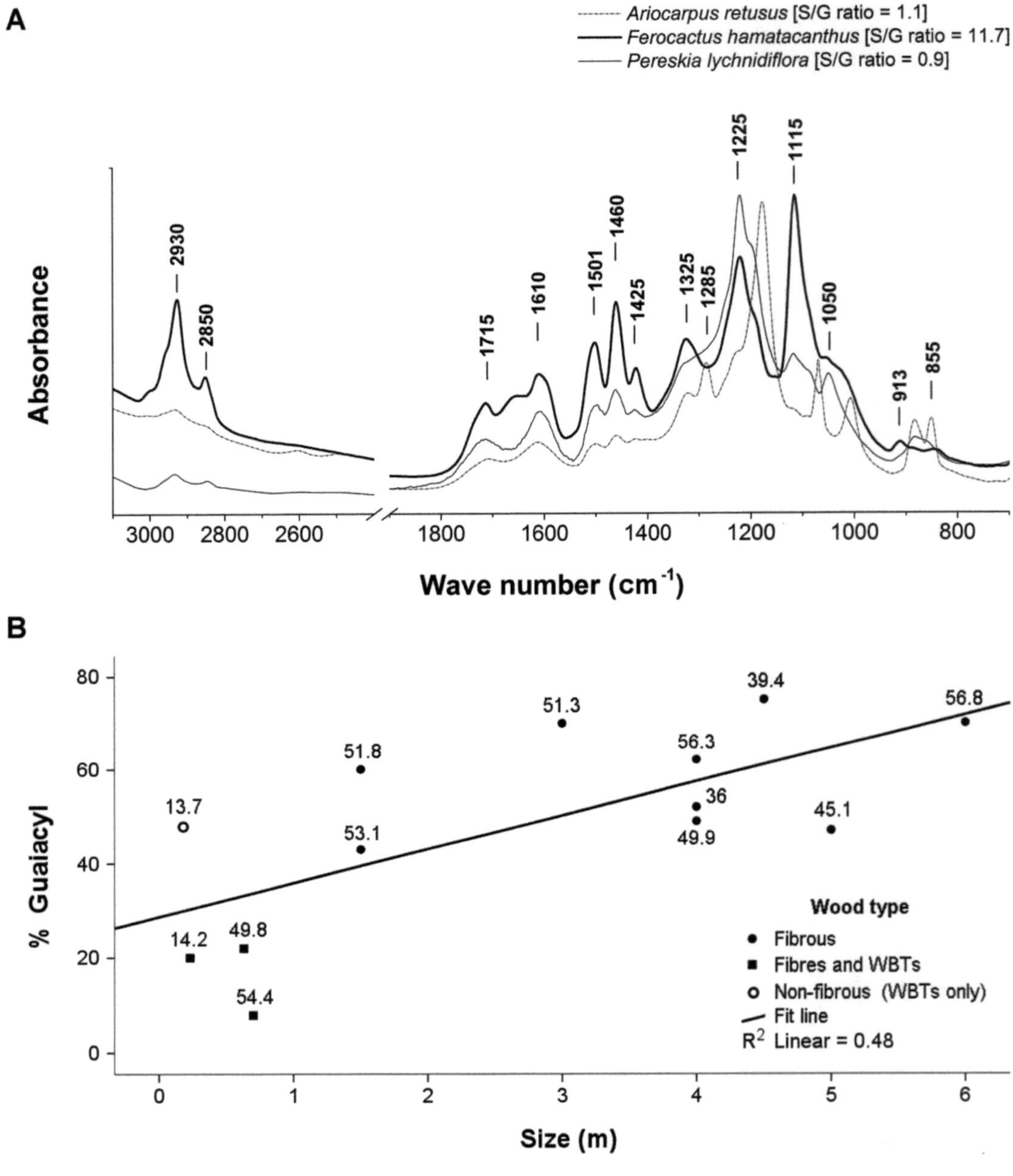


**Fig. 1** (**a**) FTIR spectra from three types of wood in Cactaceae: non-fibrous wood (*A. retusus*), wood with fibers and wide-band tracheids (WBTs) (*F. hamatacanthus*), fibrous wood (*P. lychnidiflora*). Height of the peak at 1115 $cm^{-1}$, assigned to the syringyl lignin, matches with the higher S/G ratio obtained by HPLC analysis in the *F. hamatacanthus* sample; *P. lychnidiflora* has an S/G ratio similar to other fibrous woody dicots. (**b**) Graph of the percentage of guaiacyl lignin in the wood versus size of the species in Cactaceae; all species with lignified rays and fast growth rates have more guaiacyl lignin. Numbers above the dots show the percentage of Klason lignin. Modified from Reyes-Rivera et al. 2015

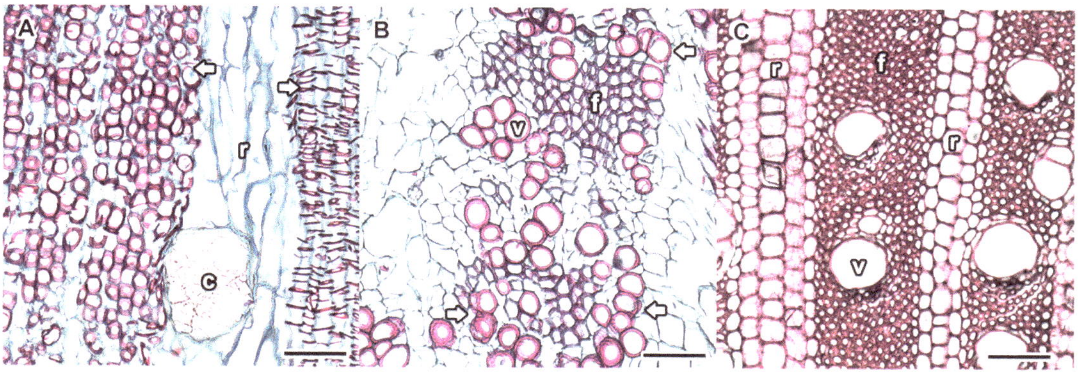


**Fig. 2** Cross sections in non-fibrous and fibrous woods in species of Cactaceae. (**a**) *Ariocarpus retusus*, (**b**) *Ferocactus hamathacanthus*, (**c**) *Pereskia lychnidiflora*. Bar is 100 μm; *c* = crystal, *r* = rays, *v* = vessels, *f* = fibers, *arrows* = wide band tracheids

5. Convection oven (e.g., catalog number PR305225M; Thermo Scientific).
6. Desiccator (e.g., catalog number 08-595E; Fisher Scientific).
7. Rotary evaporator (e.g., catalog number R 210; Buchi).
8. Cold circulator (e.g., catalog number 152-5258; Thermo Scientific).
9. Büchner funnel (e.g., catalog numbers 10-310-108, 10-358-22J; Fisher Scientific).
10. Amber glass bottles, capacity 10 mL.
11. Reflux apparatus: a stirring hotplate (SHp), an Erlenmeyer flask (EF), a magnetic stirrer bar (SB), an Allihn condenser (AC), and a cold circulator (CC). For assembly *see* Fig. 3b.
12. Vacuum filtration apparatus: a trapped vacuum pump, Kitasato flasks (KF), an Erlenmeyer flask (EF), a Büchner funnel (BF), plastic tubing (PT), a rubber stopper (RS), a crucible holder (CH), a siphon tube (ST). For assembly *see* Fig. 3c.

### 2.2 Klason Lignin

1. Sulfuric acid: 72 % $H_2SO_4$ solution, 24 N. Carefully and slowly pour 673.3 mL of concentrated $H_2SO_4$ (95.0 %, sp gr 1.84) into a 1-L volumetric flask containing 300 mL of distilled water (*see* **Note 2**), and fill the flask up to the mark. Store at room temperature.
2. Cooling bath with stirring: a magnetic stirrer (MS), a magnetic stirrer bar (SB), a crystallizing dish (CD), a volumetric flask (VF) and/or a beaker (Bk), copper tubing (CT), plastic tubing (PT), and a cold circulator (CC). For assembly *see* Fig. 3d.

### 2.3 FTIR Spectroscopy

1. Hydraulic press, 12 tons (catalog number 4386) (Carver).
2. KBr pellet die (catalog number Z506699) (Aldrich).

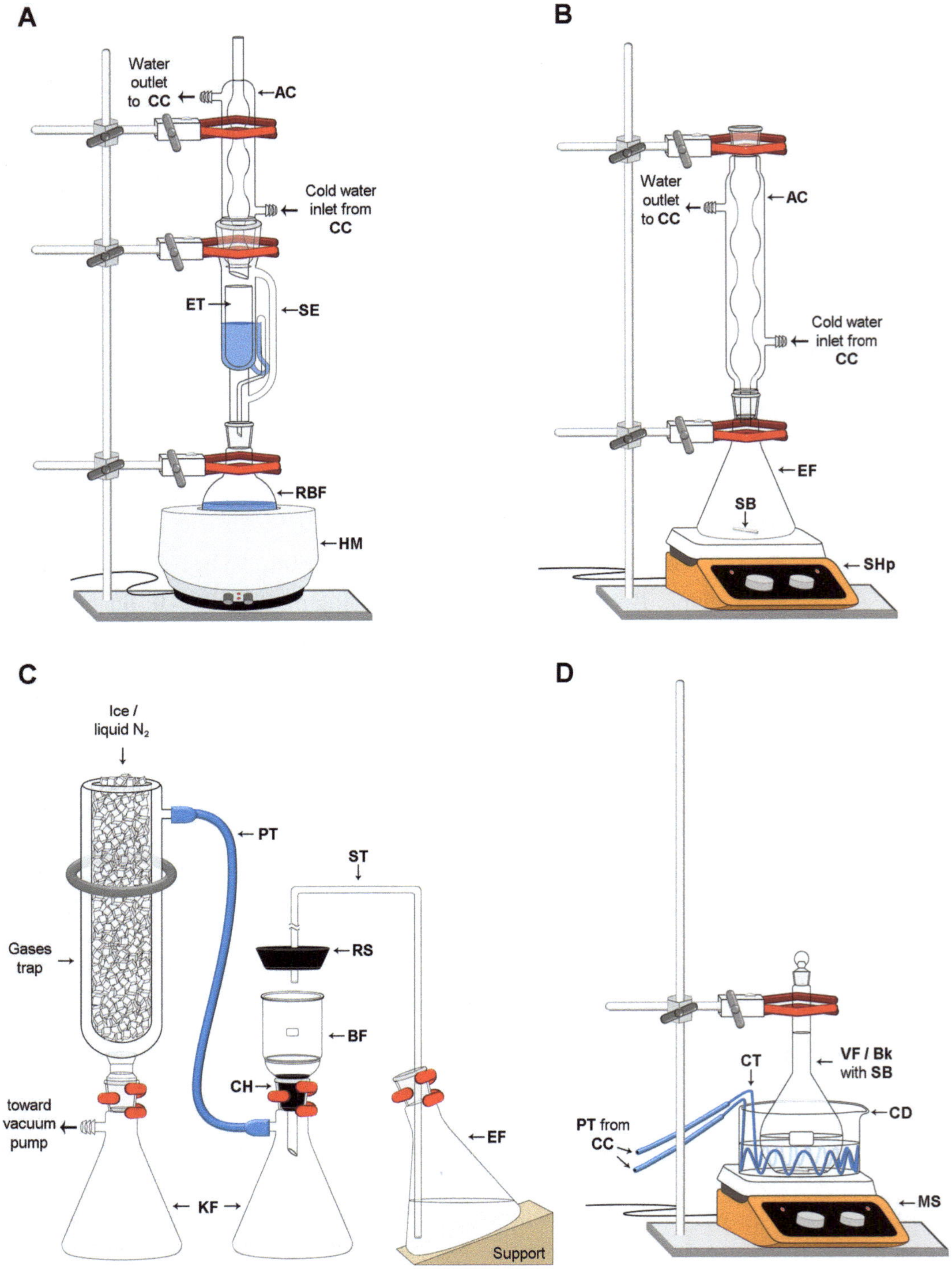


**Fig. 3** Assembling of the laboratory equipment. (**a**) Soxhlet apparatus: assembly preferably in a fume hood or in a place equipped with fume extractor. (**b**) Reflux apparatus. (**c**) Vacuum filtration apparatus. (**d**) Cooling bath with stirring, for preparation of acidic or basic aqueous solutions (use a volumetric flask), and for determining Klason lignin (use a beaker). The magnetic stirrer, crystallizer dish, and magnetic stirrer bar can be replaced by an ultrasonic bath

3. Kimwipes® (Kimtech Science).
4. Anhydrous acetone.
5. Agate mortar and pestle (Kimtech Science).
6. Potassium bromide (KBr) (catalog number 221864) (Sigma-Aldrich).
7. FTIR spectrometer (Nicolet™ series) (Thermo Scientific).
8. OMNIC™ Series Software (Thermo Scientific).

### 2.4 Alkaline Nitrobenzene Oxidation of Lignin

1. Reactor system, 2 L capacity (catalog number 4526) (Parr Instrument Company).
2. Tubular glass vials, 16 mL capacity.
3. Separatory funnels, 50 mL capacity.
4. Sodium hydroxide: NaOH solution, 2 M. Carefully and slowly transfer 81.05 g of sodium hydroxide (98.7%, sp gr 2.08) into a beaker containing 250 mL of distilled water (*see* **Note 2**) and dissolve. Pour the solution into a 1 L volumetric flask and fill the flask up to the mark. Store at room temperature until use.
5. Nitrobenzene ≥99.0% (Sigma-Aldrich).
6. Chloroform ≥99.5% (Sigma-Aldrich).
7. Hydrochloric acid: HCl solution, 4 M. Carefully and slowly pour 328.5 mL of HCl (37.0%, sp gr 1.2) into a 1-L volumetric flask containing 250 mL of distilled water (*see* **Note 2**), and fill the flask up to the mark. Store at room temperature until use.
8. pH Meter.

### 2.5 Determination of the S/G Ratio by HPLC

1. Supelco® mobile phase filtration apparatus (catalog number 58062-U) (Supelco).
2. Ultrasonic bath, capacity 2.8 L.
3. Acetonitrile-water mixture (1:1): mix together 1 volume of acetonitrile and 1 volume of water (*see* **Note 3**).
4. Acetonitrile-water mixture (1:6): mix together 1 volume of acetonitrile and 6 volumes of water (*see* **Note 3**).
5. MF-Millipore cellulose membrane, 0.22 μm (catalog number GSWP01300) (Merck Millipore).
6. Whatman® regenerated cellulose membrane filters, 0.45 μm (catalog number WHA10410212) (Sigma-Aldrich).
7. Trifluoroacetic acid buffer: 0.1% TFA solution in distilled water (*see* **Note 3**).
8. Vanillin standard (catalog number 818718) (Merck Millipore).
9. Syringaldehyde standard (catalog number 818353) (Merck Millipore).

10. Hibar® 250-4 LiChrosorb® RP-18 (5 μm) HPLC column (catalog number 150333) (Merck Millipore).
11. Manu-Cart® NT Cartridge Holder (catalog number 151486) (Merck Millipore).

## 3 Methods

Process the samples in triplicate. Record the weight of the Büchner funnels, extraction thimbles and samples until they have reached a constant weight (moisture-free). Record each weight to the nearest 0.1 mg (e.g., 1.0001 g).

### 3.1 Sample Preparation

1. Obtain a wooden disk that is 10-cm thick (*see* **Note 4**). Use a chipper or hand pruners to obtain wood chips of medium size (2 cm). Dry the wood chips in a convection oven at 50 °C for 2–4 days (*see* **Note 5**).
2. When wood chips are dry, reduce the particle size using a Wiley mill and sieving the sample through a 40- or 60-mesh screen (*see* **Note 6**).
3. Dry the milled wood in a convection oven at 105 °C to a constant weight (*see* **Note 7**).

### 3.2 Obtaining Extract-Free Wood

1. Prepare the Soxhlet apparatus (*see* Fig. 3a).
2. Weigh out 2 g of the sample and record the weight (*see* **Note 8**).
3. Weigh the extraction thimble and record its weight (*see* **Notes 7** and **8**). Place the sample in the extraction thimble (*see* **Note 9**). Label each thimble with a graphite pencil and assemble the Soxhlet apparatus (*see* Fig. 3a *and* **Note 10**).
4. Perform the Soxhlet extraction with the ethanol–benzene mixture (1:2) for 4–5 h (not less than 24 cycles). After completing the required number of cycles, dry the extraction thimble with the sample and record the weight (*see* **Note 11**). While the extraction thimble is drying, concentrate the extract (approximately 2 mL) contained in the distillation flask using a rotary evaporator (*see* **Note 12**) and store at 10 °C.
5. Repeat the extraction step as above but with a solution of ethanol (96%) as the solvent. When finished, dry the cartridge with the sample and record the weight (*see* **Note 13**). Concentrate the extract and store at 10 °C (*see* **Note 12**).
6. Prepare the reflux apparatus (*see* Fig. 3b).
7. Place the dry wood in a 1 L flask, add 850 mL of distilled water, heat to boiling and allow the sample to reflux with stirring for 1 h (*see* Fig. 3b).

8. Filter the sample through a Büchner funnel with a medium-pore fritted disk (*see* Fig. 3c *and* **Note 14**). Recover and store the extract (*see* **Note 15**).
9. Dry the Büchner funnel with the sample at 105 °C to a constant weight and record its weight (*see* **Note 8**).
10. Finally, calculate the percentage of wood extract according to the weight lost in each extraction (*see* **Note 16**) using the following formula:

$$\% = \left[ \left( A + B + C \right) / W_0 \right] \times 100$$

where:

$A$ = weight lost (g) after extraction with the ethanol–benzene mixture (2:1).

$B$ = weight lost (g) after extraction with ethanol (96%).

$C$ = weight lost (g) after extraction with hot water.

$W_0$ = initial sample weight (g).

### 3.3 Klason Lignin

1. Prepare the cooling bath with stirring at a temperature of 2 °C (*see* Fig. 3d).
2. Place 1 g of extract-free wood (record the weight to the nearest 0.1 mg; *see* **Note 8**) in a 50-mL beaker and slowly add 15 mL of 72% sulfuric acid (previously cooled at 15 °C).
3. After the acid is added, homogenize the sample with the aid of a glass stir rod. When the sample is completely impregnated with sulfuric acid the bath temperature can be raised to 18 °C and then, with the aid of tweezers, place a magnetic stir bar in the beaker and turn on the magnetic stirrer at a low speed. Cover the beaker with a watch glass and keep it in the cooling bath for 2 h (*see* **Note 17**).
4. Transfer the material to an Erlenmeyer flask containing 300 mL of distilled water, homogenize the mixture and then add another 260 mL of distilled water (560 mL of water in total).
5. Boil the solution in the reflux apparatus for 4 h with stirring (*see* **Note 18**).
6. Allow lignin to settle to the bottom of the flask while keeping it in a tilted position (*see* **Note 19**).
7. Siphon or decant the supernatant without stirring up the precipitate. Filter through a Büchner funnel with a fine-pore fritted disk that has previously been dried to a constant weight (*see* **Note 8**).
8. Transfer the precipitated lignin directly to the Büchner filter with the aid of a rod with rubber policeman.

9. Wash the lignin fraction using 50–100 mL of distilled water, then filter with just enough vacuum to remove the excess of water.
10. Dry the Büchner filter with the sample at 105 °C, allow it to cool and record the weight (*see* **Notes 8** and **20**).
11. Finally, determine the percentage of lignin with the following formula:

$$\text{Lignin}\,\% = \left(W_L / W_W\right) \times 100$$

where:

$W_L$ = lignin weight (g).

$W_W$ = weight (g) of the extractive-free wood sample.

### 3.4 Preparation of 1 % KBr Pellets for FTIR Spectroscopy

1. Prepare the hydraulic press and KBr pellet die (*see* **Note 21**).
2. Weigh out 120 mg of dry KBr (*see* **Note 22**) and place in an agate mortar (*see* **Note 23**).
3. Weigh out 1.2 mg of dry Klason lignin (*see* **Notes 8** and **24**).
4. Grind the lignin sample and the KBr until the KBr is uniform in color (*see* **Note 25**).
5. Transfer the mixture to the pellet die, place the remaining pieces (the upper anvil and the plunger) on top and apply light pressure to the plunger with the press to remove any air from the sample.
6. Increase the pressure of the hydraulic press to 11,000 pounds and maintain the pressure for 2 min.
7. Slowly release the pressure from the press and remove the die.
8. Carefully disassemble the pellet die. First remove the base and gently push the plunger to release the KBr pellet and place in a clean weighing dish (previously marked with the sample data). Place the dish in a desiccator until the pellet is ready for analysis in the FTIR spectrometer.
9. Measure the absorbance of the KBr pellets using an FTIR spectrometer. The recommended resolution is 2–4 $cm^{-1}$, in the range between 400 and 1850–4000 $cm^{-1}$, for 50–150 scans.
10. Normalize and analyze the obtained spectral data using scientific graphing and data analysis software, such as OMNIC™ (*see* **Notes 26** and **27**).

### 3.5 Alkaline Nitrobenzene Oxidation and S/G Ratio Determination by HPLC

1. Weigh out 0.2 g of dry extract-free wood (*see* **Note 8**). Transfer the sample to a tubular glass vial containing 7 mL of a 2 M NaOH aqueous solution and 0.5 mL of nitrobenzene (*see* **Note 28**).
2. Place the vial with the material (*see* **Note 29**) into a pressurized reactor in an oil bath at 170 °C for 2.5 h.

3. Filter the resulting material through a 0.22 μm Millipore cellulose membrane.
4. Transfer the filtrate to a 50-mL separatory funnel. Extract three times, using 30 mL of chloroform each time (*see* **Notes 1** and **30**).
5. Acidify the aqueous phase to pH 1 with a 4 M HCl solution. Then, extract three times (*see* **Note 30**).
6. Mix the organic phases resulting from the six extractions and concentrate the mix in a rotary evaporator at 40 °C under reduced pressure.
7. Transfer the sample to a 50-mL volumetric flask and fill the flask up to the mark with acetonitrile-water (1:1).
8. Filter the sample through a regenerated cellulose membrane with a pore size of 0.45 μm.
9. Analyze the sample on an HPLC system, preferably one equipped with a variable wavelength or UV-VIS detector, an analytical reverse phase column RP-18 type (250×4.6 mm, 5 μm), and a standard flow cell, at a wavelength of 280 nm with an isocratic elution. Use the acetonitrile/water mixture (1:6) as the mobile phase and adjust to pH 2.6 using the 0.1 % trifluoroacetic acid buffer solution in distilled water.
10. Make calibration curves using syringaldehyde and vanillin standards. Prepare the standards using the mixture of acetonitrile/water (1:6) and adjust the pH to 2.6 (*see* **Note 31**).
11. Finally, calculate the S/G ratio (*see* **Note 32**).

## 4 Notes

1. Special care is recommended when handling solvents such as benzene, chloroform and other agents used in the methods described herein. These have been identified as highly carcinogenic agents with extended periods of exposure. Read the safety data sheets of each reagent before handling.
2. Because the preparation of acidic or basic aqueous solutions is a highly exothermic process, it is recommended to prepare these solutions in a cold bath at 15 °C (*see* Fig. 3d). Never add water directly to acids or bases because they may release enough heat to cause severe burns.
3. To prevent damage to the HPLC apparatus and column, filter all solvents using a mobile phase filtration apparatus and a regenerated cellulose membrane, pore size 0.45 μm. Furthermore, degas each solvent prior to use, either by a supersonic bath, evacuation with a vacuum pump, or with bubbling helium. The time of each degassing method depends on the sample volume.

4. The age of the wood varies according to the anatomical region of the stem. Cambial maturation occurs radially: the juvenile wood is in the innermost zone of the cylinder, while the mature wood is in the outermost region. Additionally, the upper wood of the stem can be considered relatively younger than the wood in the base of the stem. Thus, a complete ontogenetic history of the wood in an individual is represented in the wood from the base of the stem.
5. The drying time of the wood depends on the degree of succulence of the sample. For example, for cacti, it is necessary to dry samples for 4 days and store them in a dry and well-ventilated place to prevent infestation by fungi.
6. When the determination of Klason lignin is performed with a small amount of sample (<0.5 g), the homogenization of the sample with the sulfuric acid solution is most effective when working with a particle size of 0.25 mm (mesh 60).
7. It is recommended to simultaneously dry the extraction thimbles and wood samples until each reach a constant weight.
8. Always record the weight to the nearest 0.1 mg. Make sure each object (samples, extraction thimbles or Büchner funnels) have reached a constant weight while drying at 105 °C and only weigh them when they are humidity-free. For this, they can be placed in a desiccator while being transported from the convection oven to the analytical balance. This will also allow the objects to cool. To avoid burns, handle hot objects (e.g., Büchner funnels) with crucible tongs and asbestos gloves.
9. Fold the open end of the extraction thimble to prevent sample loss during reflux.
10. The volume of solvent in the distillation flask should contain two to three times the volume of the Soxhlet extractor to ensure the continuous flow of solvent.
11. Air dry until the solvent has evaporated, then dry at 105 °C to a constant weight and record the weight of the extraction thimble with the sample.
12. Transfer the extract to an amber glass bottle and recover the extract remaining in the distillation flask with a small amount of clean solvent. The extracts may be used to characterize the solvent extractives of the wood later.
13. Air dry until the solvent has evaporated, then dry at 105 °C to a constant weight and record the weight of the extraction thimble with the sample. Discard the extraction thimble and then record the weight (constant weight) of the sample only.
14. Dry the Büchner funnel (medium pore-size), before starting the process, at 105 °C to a constant weight and record its weight.

This will be used to calculate the weight of the soluble extractives in the hot water. The filtering system is shown in Fig. 3c.

15. This fraction contains substances such as tannins, gums, sugars, starches and coloring matter. If analyzing this fraction is required, we recommend that it be preserved by freeze-drying to prevent infestation by fungi.
16. It is important to calculate the weight of each solvent extractive fraction independently of the weight of the extraction thimble or Büchner funnel. (a) After extraction with ethanol–benzene (1:2), subtract the weight of the extraction thimble from the weight of the extraction thimble with the sample and use the resulting sample weight (ethanol–benzene extractive-free wood) to calculate the ethanol–benzene extractives (weight lost after extraction with the ethanol–benzene mixture) with respect to the initial weight of the sample. (b) After extraction with ethanol (96%), subtract the weight of the extraction thimble with the sample from the weight of the extraction thimble with the sample obtained in the extraction with ethanol–benzene. Use the resulting weight to calculate the weight of the extractives soluble in ethanol (weight lost after the extraction with ethanol) with respect to the weight of the sample after extraction with the ethanol–benzene mixture. (c) After extraction with hot water, subtract the weight of the porous filter from the weight of the porous filter with the sample. Use the resulting sample weight to calculate the weight of extractives soluble in hot water with respect to the weight of the sample obtained in the ethanol extraction.
17. During the agitation make sure that the sample does not adhere to the beaker walls. If that happens, use the glass rod to remove the sample and add it back to the solution.
18. It is not recommended to use the reflux apparatus if the content of lignin-soluble acid in the solution will be determined separately. If this is the case, keep the volume constant by adding water continuously instead of refluxing the solution.
19. You can let the sample stand overnight because the settling time can take several hours.
20. The weight of the lignin is obtained by subtracting the initial weight of the Büchner funnel (without the sample) from the weight of the Büchner funnel with the sample.
21. Completely clean the KBr pellet die before and after use, particularly the plunger and the anvil, to prevent sample contamination. Use dry acetone and Kimwipes® to avoid scratching the surfaces; a scratched surface affects the crystallinity of the KBr pellets. Then, reassemble the mold without replacing the upper anvil and the plunger.

22. For comparative studies, the KBr used should be of FTIR grade. In addition, the KBr should be dried previously at 105 °C to a constant weight to avoid interference in the FTIR spectra. For the normalization of FTIR spectra, the weight should be recorded to the nearest to 0.1 mg.
23. Do not use porcelain mortars because the porosity of the material can cause contamination of lignin samples.
24. In places with high humidity, it is convenient to place the mortar with the KBr into a desiccator while the lignin sample is weighed to prevent wetting the sample.
25. Grind the mixture as quickly as possible to avoid excessive hydration of the sample due to environmental humidity.
26. Bands that do not vary from one sample to another (selection of those bands will depend on the spectra used to normalize the data as a whole) are used to normalize the FTIR spectra of the lignin. It is recommended to make a pellet consisting of only KBr to use as a blank; this will help in discarding bands produced by noise or bands that are foreign to the sample in study.
27. Bands 1501–1505 $cm^{-1}$ are commonly used to determine the lignin content based on FTIR spectra. The filtering system is shown in Fig. 3c. To normalize the bands and make them quantitatively comparable, the height of the band observed in the FTIR spectrum is divided by the concentration of lignin in the KBr pellet [9, 38, 47, 48].
28. First, shake the vial with the mixture of NaOH and nitrobenzene so that the nitrobenzene is homogeneously distributed within the mixture, then add the sample of dry wood and ensure that it is homogeneously impregnated.
29. Place a cotton pad over the top of the vial and put the screw cap on firmly so that the cotton pad is well secured between the vial and the cap.
30. Constantly purge the gas formed within the separatory funnel in each extraction. The amount of gas formed at the beginning of each extraction, due to the use of new solvent or acidification, can be considerable and can cause the funnel to open, resulting in sample loss. Work in a fume hood and wear a gas-filtering mask, protective gloves and eye protection.
31. As a reference for construction of the calibration curves, the following previously used concentrations used by Nunes et al. [28] are given: for vanillin, 0.375, 0.75, 1.125, and 1.5 mmol/L; for syringaldehyde, 0.825, 1.65, 2.475, and 3.3 mmol/L.
32. For the interpretation of the results, it should be taken into account that, due to their high proportion of condensation,

approximately 30% of the guaiacyl units are converted to vanillin. Conversely, due to their low proportion of condensation, 90% of syringyl units are converted to syringaldehyde.

## References


1. Sjöström E (1981) Wood chemistry, fundamentals and applications. Academic, New York
2. Hon DN, Shiraishi N (eds) (2001) Wood and cellulosic chemistry. Marcel Dekker, New York
3. Koch G (2006) Raw material for pulp. In: Sixta H (ed) Handbook of pulp. Wiley-VCH, Weinheim
4. Sarkanen KV, Ludwig CH (eds) (1971) Lignins: occurrence, formation, structure and reactions. Wiley-Interscience, New York
5. Lin SY, Dence CW (eds) (1992) Methods in lignin chemistry. Springer, Berlin
6. Sakakibara A, Sano Y (2001) Chemistry of lignin. In: Shirashi N, Hon DN (eds) Wood and cellulosic chemistry. Marcel Dekker, New York
7. Lupoi JS, Singh S, Parthasarathi R et al (2015) Recent innovations in analytical methods for the qualitative and quantitative assessment of lignin. Renew Sustain Energy Rev 49:871–906
8. Colom X, Carrillo F, Nogués F, Garriga P (2003) Structural analysis of photodegraded wood by means of FTIR spectroscopy. Polym Degrad Stab 80:543–549
9. Pandey KK, Pitman AJ (2003) FTIR studies of the changes in wood chemistry following decay by brown-rot and white-rot fungi. Int Biodeterior Biodegradation 52:151–160
10. Nuopponen M, Vuorinen T, Jaemsae S, Viitaniemi P (2004) Thermal modifications in softwood studied by FT-IR and UV resonance Raman spectroscopies. J Wood Chem Technol 24:13–26
11. Colom X, Carrillo F (2005) Comparative study of wood samples of the northern area of Catalonia by FTIR. J Wood Chem Technol 25:1–11
12. Kubo S, Kadla JF (2005) Hydrogen bonding in lignin: a Fourier transform infrared model compound study. Biomacromolecules 6:2815–2821
13. Xu F, Sun J-X, Sun RC et al (2006) Comparative study of organosolv lignins from wheat straw. Ind Crops Prod 23:180–193
14. Del Rio JC, Gutierrez A, Rodriguez IM et al (2007) Composition of non-woody plant lignins and cinnamic acids by Py-GC/MS, Py/TMAH and FT-IR. J Anal Appl Pyrol 79:39–46
15. Mansouri N-E, Salvadó J (2007) Analytical methods for determining functional groups in various technical lignins. Ind Crops Prod 26:116–124
16. Müller G, Schöpper C, Vos H et al (2009) FTIR-ATR spectroscopic analysis of changes in wood properties during particle- and fibreboard production of hard- and softwood trees. Bioresources 4:49–71
17. Brebu M, Tamminen T, Spiridon I (2013) Thermal degradation of various lignins by TG-MS/FTIR and Py-GC–MS. J Anal Appl Pyrol 104:531–539
18. Hu J, Xiao R, Shen D, Zhang H (2013) Structural analysis of lignin residue from black liquor and its thermal performance in thermogravimetric-Fourier transform infrared spectroscopy. Bioresource Technol 128:633–639
19. Sammons RJ, Harper DP, Labbe N et al (2013) Characterization of organosolv lignins using thermal and FT-IR spectroscopic analysis. BioResources 8:2752–2767
20. Xu G, Wang L, Liu J, Wu J (2013) FTIR and XPS analysis of the changes in bamboo chemical structure decayed by white-rot and brown-rot fungi. Appl Surf Sci 280:799–805
21. Fonseca-Prieto FV, Canché-Escamilla G, Chavarria-Hernandez JC, Duarte-Aranda S (2014) Characterization of lignocellulosic residues of henequen and their use as a bio-oil source. Biomass Convers Bioref 4:95–104
22. Reyes-Rivera J, Canché-Escamilla G, Soto-Hernández M, Terrazas T (2015) Wood chemical composition in species of Cactaceae: the relationship between lignification and stem morphology. PLoS One 10:e0123919. doi:10.1371/journal.pone.0123919
23. Allison GG, Thain SC, Morris P et al (2008) Quantification of hydroxy cinnamic acids and ligninin perennial forage and energy grasses by Fourier-transform infrared spectroscopy and partial least squares regression. Bioresource Technol 100:1252–1261
24. Huang Y, Wang L, Chao Y et al (2012) Analysis of lignin aromatic structure in wood based on the IR spectrum. J Wood Chem Technol 32:294–303

25. Derkacheva OY (2013) Estimation of aromatic structure contents in hardwood lignins from IR absorption spectra. J Appl Spectrosc 80:670–676
26. Lupoi JS, Singh S, Davis M et al (2014) High-throughput prediction of eucalypt lignin syringyl/guaiacyl content using multivariate analysis: a comparison between mid-infrared, near-infrared, and Raman spectroscopies for model development. Biotechnol Biofuels 7:93
27. Gora R, Hutta M, Vrska M et al (2006) Characterization of Klason lignins by reversed phase high-performance liquid chromatography using wide-pore octadecyl silica and stepwise gradient of dimethyl formamide in water. J Sep Sci 29:2179–2189
28. Nunes CA, Lima CF, Barbosa LCA et al (2010) Determination of *Eucalyptus spp* lignin S/G ratio: a comparison between methods. Bioresource Technol 101:4056–4061
29. Gomes DI, Detmann E, Valadares FSC et al (2011) Evaluation of lignin contents in tropical forages using different analytical methods and their correlations with degradation of insoluble fiber. Anim Feed Sci Technol 168:206–222
30. Dence CW (1992) The determination of lignin. In: Lin SY, Dence CW (eds) Methods in lignin chemistry. Springer, Berlin
31. Ma XJ, Cao SL, Yang XF et al (2014) Lignin removal and benzene-alcohol extraction effects on lignin measurements of the hydro-thermal pretreated bamboo substrate. Bioresource Technol 151:244–248
32. Nicholson DJL, Aaron T, Francis, Raymond CA (2013) A three-stage Klason method for more accurate determinations of hardwood lignin content. Cellulose Chem Technol 48:53–59
33. Rencoret J, Gutiérrez A, Nieto L et al (2011) Lignin composition and structure in young versus adult *Eucalyptus globulus* plants. Plant Physiol 155:667–682
34. Bose SK, Francis RC, Govender M et al (2008) Lignin content versus syringyl to guaiacyl ratio amongst poplars. Bioresource Technol 100:1628–1633
35. Chen CL (1992) Nitrobenzene and cupric oxide oxidations (of ligninin solution). In: Lin SY, Dence CW (eds) Methods in lignin chemistry. Springer, Berlin
36. Sun X-F, Jing Z, Fowler P et al (2011) Structural characterization and isolation of lignin and hemicelluloses from barley straw. Ind Crops Prod 33:588–598
37. Ohra-aho T, Gomes FJ, Colodette JL, Tamminen T (2013) S/G ratio and lignin structure among *Eucalyptus* hybrids determined by Py-GC/MS and nitrobenzene oxidation. J Anal Appl Pyrol 101:166–171. doi:10.1016/j.jaap.2013.01.015
38. Rodrigues J, Faix O, Pereira H (1998) Determination of lignin content of *Eucalyptus globulus* wood using FTIR spectroscopy. Holzforschung 52:46–50
39. Schultz TP, Burns DA (1990) Rapid secondary analysis of lignocellulose: comparison of near infrared (NIR) and Fourier transform infrared (FTIR). Tappi J 73:209–212
40. Stefke B, Windeisen E, Schwanninger M, Hinterstoisser B (2008) Determination of the weight percentage gain and of the acetyl group content of acetylated wood by means of different infrared spectroscopic methods. Anal Chem 80:1272–1279
41. Toivanen TJ, Alen R (2007) A FTIR/PLS method for determining variations in the chemical composition of birch (*Betula pendula/B. publescens*) stem wood. Appita J 60:155–160
42. Dang VQ, Bhardwaj NK, Hoang V, Nguyen KL (2007) Determination of lignin content in high-yield kraft pulps using photoacoustic rapid scan Fourier transform infrared spectroscopy. Carbohydr Polym 68:489–494
43. Faix O, Bottcher JH (1993) Determination of phenolic hydroxyl group contents in milled wood lignins by FTIR spectroscopy applying partial least-squares (PLS) and principal components regression (PCR). Holzforschung 47:45–49
44. Raiskila S, Pulkkinen M, Laakso T et al (2007) FTIR spectroscopic prediction of Klason and acid soluble lignin variation in Norway spruce cutting clones. Silva Fennica 41:351–371
45. Rana R, Langenfeld-Heyser R, Finkeldey R, Polle A (2010) FTIR spectroscopy, chemical and histochemical characterisation of wood and lignin of five tropical timber wood species of the family of Dipterocarpaceae. Wood Sci Technol 44:225–242
46. Chen H, Ferrari C, Angiuli M et al (2010) Qualitative and quantitative analysis of wood samples by Fourier transform infrared spectroscopy and multivariate analysis. Carbohydr Polym 82:772–778
47. Carballo AL, Igarza OU, Cordero ME (2004) Composición química de tres maderas en la Provincia del Pinar del Río, Cuba a tres alturas del fuste comercial. Parte N° 1: *Corymbia citriodora*. Rev Chapingo S Cienc Forest Amb 10:57–62
48. Orea IU, Carballo AL, Cordero ME (2004) Composición química de tres maderas en la Provincia del Pinar del Río, Cuba a tres alturas

del fuste comercial. Parte N° 2: *Eucalyptus pellita* F. Muell. Rev Chapingo S Cienc Forest Amb 10:51–55

49. Hergert HL (1971) Infrared spectra. In: Sarkanen KV, Ludwid CH (eds) Lignins: occurrence, formation, structure and reactions. Wiley-Interscience, New York
50. Faix O (1991) Classification of lignins from different botanical origins by FTIR spectroscopy. Holzforschung 45:21–27
51. Fengel D, Wegener G (2003) Wood, chemistry and ultrastructure and reactions. Kassel Verlag, Ramagen
52. Brebu M, Cazacu G, Chirila O (2010) Pyrolysis of lignin—a potential method for obtaining chemicals and/or fuels. Cellulose Chem Technol 45:43–50
53. Derkacheva O, Sukhov D (2008) Investigation of lignins by FTIR spectroscopy. Macromol Symp 265:61–68

# Chapter 15

# Carbohydrate Composition Analysis in Xylem

**Baocai Zhang and Yihua Zhou**

## Abstract

Gas chromatography-mass spectrometry (GC-MS) is a widely applied technique in carbohydrate composition analysis due to high accuracy and repeatability. As cell walls are the major carbohydrate pool in xylem tissues, the measurement of carbohydrate composition essentially constitutes an analysis of wall residues. Here, we describe the method for quantification of the contents of several common neutral monosaccharides and cellulose, the major components of cell walls in poplar xylem tissues, by GC-MS. The procedures for extraction of cell wall residues from xylem, generation of alditol derivatives, subsequent further separation on gas chromatograph, and detection by mass spectrometer are included.

**Key words** Cellulose, Monosaccharide, Alditol acetate, Gas chromatography, Mass spectrometry, Neutral sugar

## 1 Introduction

Xylem, which transports water and nutrients, is a well-characterized tissue in vascular plants due to the presence of thickened secondary walls [1]. Cell walls are a complex polysaccharide network, composed of cellulose microfibrils and a matrix of hemicelluloses, pectins, lignins, and glycoproteins [2–4]. Therefore, determining the carbohydrate composition is essential for revealing the physiochemical properties of xylem walls. Colorimetric methods and paper chromatography act as preliminary methods to examine sugar content [5–8]. However, these methods cannot quantify the carbohydrate composition with high sensitivity. The gas chromatography (GC) method has become popular since Peter Albersheim and his colleagues set up a method for generation of alditol acetate derivatives from hydrolyzed monosaccharides [9, 10], and in particular when a flame-ionization detector was replaced by mass spectrometer [11, 12]. The method described here is complementary to that for producing trimethylsilyl ethers of methyl glycosides for the analysis of uronic acid content [10, 13]. Several other techniques, such as high performance liquid chromatography [14, 15]

Miguel de Lucas and J. Peter Etchells (eds.), *Xylem: Methods and Protocols*, Methods in Molecular Biology, vol. 1544,
DOI 10.1007/978-1-4939-6722-3_15, 

and nuclear magnetic resonance spectroscopy [16], are also useful and sometimes are more suitable to determine the content or structure of individual polysaccharide. Nevertheless, due to advantages in sensitivity, comprehensiveness, and stability, gas chromatography-mass spectrometry (GC-MS) is the most widely used method for analysis of monosaccharide composition. Here, we describe the procedures to extract wall residues from wood xylem tissues, to hydrolyze wall residues, to generate and separate the alditol derivatives on gas chromatograph, and to detect them by mass spectrometer. The methods for examination of cellulose content and quantification of carbohydrate are referenced to the previous reports [7, 9–11] with revisions.

## 2 Materials

### 2.1 Reagents

Prepare all solutions using chromatographic grade reagents and ultrapure water (18 MΩ cm at 25 °C). Prepare and store all reagents at room temperature. Strictly follow all waste disposal regulations when disposing of waste materials. Sodium azide is only used in the destarch steps.

1. Chloroform/methanol solution: Mix 250 ml of chloroform and 250 ml of methanol.
2. Sodium acetate buffer: 0.1 M sodium acetate in water, pH 5.0. Weigh 4.10 g sodium acetate and transfer to a beaker, add water to a volume of 450 ml. Mix and adjust pH with acetic acid. Make up to a volume of 500 ml with water.
3. Inositol solution: Weigh 20 mg of inositol and dissolve in 10 ml water. Store aliquots at −20 °C.
4. Sodium borohydride solution: Dissolve 150 mg of sodium borohydride in 15 ml of ammonium hydroxide (1 M). Prepare freshly prior to use.
5. Acetic acid-methanol solution: Mix 50 ml of glacial acetic acid in 450 ml of methanol.
6. Trifluoroacetic acid (TFA) solution: 2 M TFA in water. Dilute 15.41 ml of TFA (Sigma, St. Louis, MO, USA) to 100 ml with water in fume hood.
7. Acetic nitric reagent (8:1:2 Acetic acid:nitric acid:water): Mix 30 ml of concentrated nitric acid in 300 ml of 80% acetic acid.
8. Sulfuric acid solution: 72% sulfuric acid in water. Slowly add 65.2 ml of 98% concentrated sulfuric acid into 34.8 ml water with stirring.
9. Anthrone reagent: 0.2% Anthrone in concentrated sulfuric acid. Weigh 40 mg of anthrone and transfer to a flask containing 20 ml of concentrated sulfuric acid. Chill in a refrigerator for 2 h. Prepare freshly prior to use.

10. Sodium azide solution: 0.01 % Sodium azide in water.
11. Amylase solution: Weigh 5 mg of amylase (*Bacillus* species, Sigma, St. Louis, MO, USA) and dissolve in 20 ml water. Store aliquots at −20 °C.
12. Pullulanase M1 solution (*Klebsiella planticola*, Megazyme, Bray, Ireland). Store at 4 °C.
13. Iodine-postassium iodide solution: Weigh 0.4 g of potassium iodide and dissolve in 60 ml of water, then add 0.2 g of iodine and mix well.

### 2.2 Equipment

1. Tissuelyser II (Qiagen, CA, USA, or an equivalent mill) and stainless-steel tubes and beads.
2. Freeze-drier.
3. Screw-capped glass tubes (8 ml).
4. Vortex.
5. Centrifuge 5810R (Eppendorf, Hamburg, Germany) or an equivalent benchtop centrifuge.
6. Digital scale with the resolution at 0.01 mg.
7. Dri-Block equipped with a sample concentrator (Techne, Cambridge, UK) or an equivalent concentrator.
8. GC-MS system (e.g., Agilent 7890 GC with an Agilent 5977 MS instrument, Agilent, Santa Clara, CA, USA).
9. GC column (SP-2380, 30 m × 0.25 mm inner diameter × 0.2 μm film thickness fused silica capillary column, Supelco, Bellefonte, PA, USA).
10. High-purity helium (>99.999 %).
11. GC Vials (2 ml) with inserts (400 μl) and Polytetrafluoroethylene (PTFE)-lined caps.
12. pH monitor.
13. Positive displacement pipette, e.g., Microman (Gilson, Villiers-le-Bel, France).

## 3 Methods

Carry out all procedures (Fig. 1) at room temperature unless otherwise specified. All the evaporation steps should be performed in fume hood.

### 3.1 Alcohol Insoluble Residue Preparation

1. Cut the stems, peel off the bark, and harvest the xylem for cell-wall residue preparation.
2. Freeze dry in lyophilizer (optionally air dry in oven).

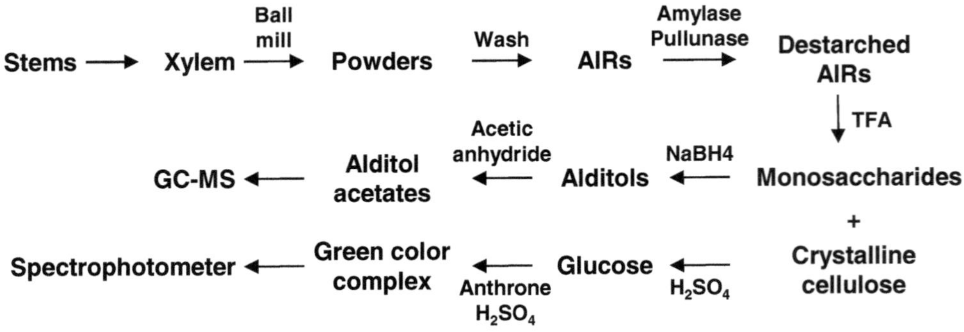


**Fig. 1** Overview of the major procedure for carbohydrate analysis. Xylem tissues are collected from the target stems and ball milled into powders. The alcohol insoluble residues (AIRs) are prepared by successively rinsing the samples in aqueous ethanol, chloroform-methanol, and acetone solutions, and destarching with amylase and pullunase. The destarched AIRs are further hydrolyzed into monosaccharides and reduced with sodium borohydride. The alditol acetate derivatives produced by acetic anhydride treatment are subjected to GC-MS analysis. The remains of TFA treatment are hydrolyzed with sulfuric acid. After development of color with anthrone in concentrated sulfuric acid, the absorption of each reaction at 625 nm is measured on a spectrophotometer. The contents of monosaccharide and cellulose are determined based on the sugar standard curves

3. Ball mill into powder in stainless steel tubes using Tissuelyser II (Qiagen, CA, USA) with the setting: 1 min, 30 Hz and pass through 100 mesh sieve.
4. Weigh 100 mg powder into 7 ml centrifuge tubes.
5. Add 4 ml of 70% aqueous ethanol and centrifuge at 1,300 × *g* for 10 min to collect the pellet.
6. Wash the pellet with 4 ml of chloroform/methanol solution and centrifuge at 1,300 × *g* for 10 min to collect the pellet.
7. Wash the pellet with 500 μl of acetone and centrifuge at 1,300 × *g* for 10 min to collect the pellet.
8. Dry the pellets in an oven at 42 °C to obtain the alcohol insoluble residues (AIRs).
9. Resuspend the AIRs in 5 ml of sodium acetate buffer (*see* **Note 1**). Heat at 80 °C for 20 min to inactivate the endogenous enzymes and cool on ice.
10. Successively add 5 μl of sodium azide solution, 20 μl of amylase solution (7.5 U) and 5 μl of pullulanase solution (3.5 U), and incubate at 37 °C overnight to remove starch from AIRs (*see* **Note 2**).
11. Heat at 100 °C for 10 min to stop the reaction.
12. Centrifuge at 1,300 × *g* for 10 min to pellet the destarched AIRs (*see* **Note 1**).
13. Wash three times with water and centrifuge at 1,300 × *g* for 10 min to collect AIRs in each wash.

14. Freeze dry the AIRs in a lyophilizer (optionally further wash with acetone and air dry in an oven at 42 °C instead of lyophilization).

### 3.2 Acid Hydrolysis

1. Weigh 2 mg of destarched AIR powder into 2 ml centrifuge tubes and add 20 μl of inositol solution (40 μg of inositol) as an internal standard (*see* **Note 3**).
2. Wash the tube wall with 200 μl of acetone and air dry in a heating block or an oven at 42 °C (*see* **Note 4**).
3. Add 250 μl of TFA solution, cap tightly, and incubate at 121 °C for 90 min in the heat block (Dri-Block DB200, Techne, Cambridge, UK) to hydrolyze the matrix polysaccharides into monosaccharides (*see* **Note 5**).
4. Cool the samples by transferring the heat block on ice. Centrifuge the samples at 11,000 × *g* for 10 min.
5. Transfer 200 μl of supernatant into glass tubes with screw cap. Evaporate the supernatant with a gentle air stream. The remains of hydrolysis are kept for cellulose assays.
6. Add 300 μl of isopropanol into the glass tubes from **step 5** and evaporate under a gentle air stream at 42 °C to remove the remaining TFA.
7. Repeat the above **step 6** for two more times.

### 3.3 Monosaccharide Derivation

1. Add 200 μl of sodium borohydride solution to each glass tube, cap tightly, and incubate at room temperature for 90 min to reduce the monosaccharides into alditols.
2. Add 150 μl of glacial acetic acid to convert the excess sodium borohydride into borate.
3. Add 250 μl of acetic acid-methanol solution and evaporate under a gentle air stream at 42 °C.
4. Repeat the above **step 3** for two times (*see* **Note 6**).
5. Add 250 μl of methanol solution and evaporate under a gentle air stream at 42 °C.
6. Repeat the above **step 5** for two times (*see* **Note 6**).
7. Add 50 μl of acetic anhydride and 50 μl of pyridine, cap tightly, and incubate at 121 °C for 20 min to generate alditol acetate derivatives (*see* **Note** 7).
8. Cool the heat block containing the samples on ice. Uncap and evaporate under a gentle air stream at 42 °C.
9. Add 200 μl of toluene and evaporate under a gentle air stream at 42 °C.
10. Repeat the above **step 9** two times.
11. Add 1 ml of ethyl acetate and vortex.

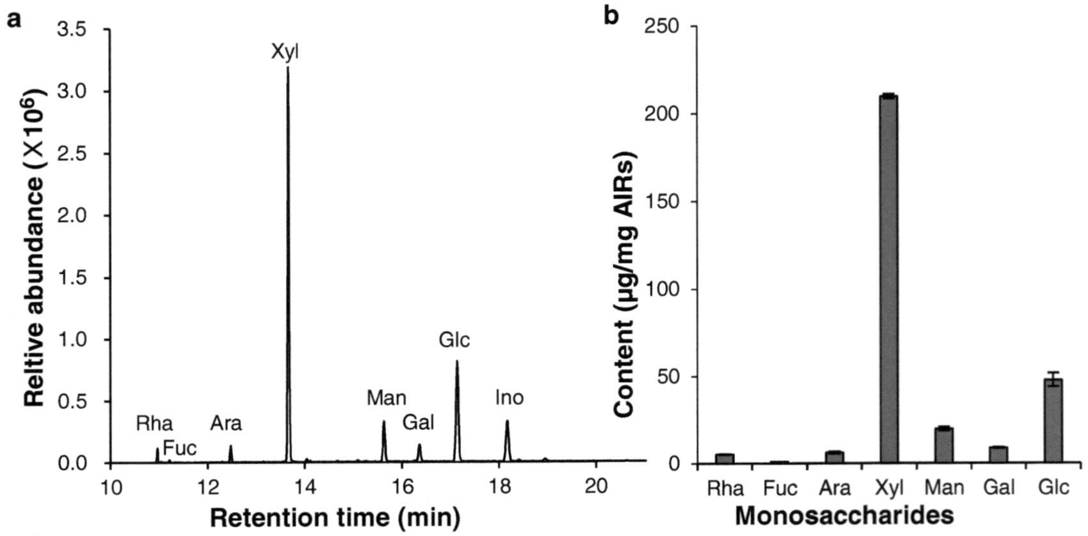


**Fig. 2** Composition analysis of matrix polysaccharides in poplar xylem. (**a**) Chromatograph of alditol acetates derived from one biological replicate of *Populus trichocarpa* xylem using the described protocol. (**b**) The content of neutral monosaccharides in *Populus trichocarpa* xylem. Error bars indicate the standard error of four biological replicates. *Fuc* fucose, *Rha* rhamnose, *Ara* arabinose, *Xyl* xylose, *Man* mannose, *Gal* galactose, *Glc* glucose, *Ino* inositol (an internal standard)

12. Add 4 ml of water, cap tightly and vortex.
13. Centrifuge at 600 × *g* for 2 min to partition per-*O*-acetylated alditols into ethyl acetate layer.
14. Transfer 100 μl of ethyl acetate layer solutions (upper phase) into the inserts in gas chromatography vials and supplement with 200 μl of acetone for dilution.

### *3.4 Gas Chromatography and Mass Spectrometry Analysis*

1. Inject 2 μl of the samples prepared in Subheading 3.3 into gas chromatography columns (SP-2380) in split mode (ratio 20:1) using helium as carrier gas constantly at 1 ml/min.
2. Separate per-*O*-acetylated alditols under the oven ramp program: 160 °C for 2 min; ramp to 200 °C at 20 °C/min and hold for 5 min; ramp to 245 °C at 20 °C/min and hold for 12 min. Maintain the mass selective detector (MSD) transfer line at 280 °C.
3. Acquire the mass spectrum data in scan mode (50–500 $m/z$) and with a 6 min solvent delay. Maintain the MS quadrupole at 150 °C and the ion source at 230 °C. An example of total ion chromatograph is shown (Fig. 2a).
4. Identify the mass peaks of each monosaccharide based on the mass profile and the retention time of corresponding standards (*see* **Note 8**).
5. Quantify the content of monosaccharides (Fig. 2b) based on the standard curves with the internal inositol control (*see* **Note 9**).

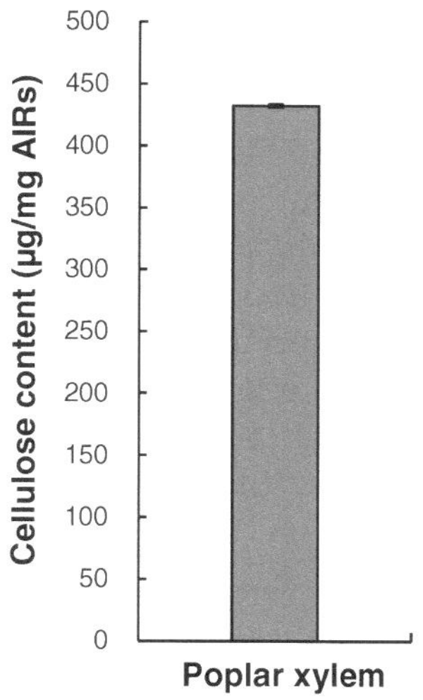


**Fig. 3** The cellulose content in poplar xylem. The cellulose content is determined by the anthrone assay using the described protocol. Error bars indicate the standard error of four biological replicates

### 3.5 Cellulose Determination

1. Add 1 ml of acetic nitric reagent to the pellets obtained from **step 5** in Subheading 3.2, and mix well by vortex.
2. Incubate at 100 °C for 30 min, and cool samples on ice.
3. Centrifuge at 11,000 × *g* for 10 min and discard the supernatant immediately.
4. Wash the pellets with water and centrifuge at 11,000 × *g* for 10 min.
5. Wash the pellets with acetone for four times and centrifuge at 11,000 × *g* for 10 min in each wash.
6. Air dry the pellets in the heat block or in the oven at 42 °C (*see* **Note 4**).
7. Add 175 μl of sulfuric acid solution (72%) with positive displacement pipette (*see* **Note 10**) and incubate at room temperature for 1 h.
8. Add 825 μl of water and vortex.
9. Centrifuge at 11,000 × *g* for 5 min.
10. Transfer 10 μl of supernatant and 90 μl of water into the wells of a microtiter plate.
11. Add 200 μl of anthrone reagent with positive displacement pipette (*see* **Note 10**), heat at 80 °C for 0.5 h for fully color development, and cool at room temperature.
12. Read the absorption of each reaction at 625 nm on a spectrophotometer against a reagent blank within 1 h (*see* **Note 11**).

13. Quantify the content (Fig. 3) based on the standard curve of glucose (*see* **Note 12**).

This method for monosaccharide composition and cellulose content analyses could be applied to handle from several to tens of samples once. To improve the efficiency of measurement, we prefer to conduct heating and evaporating steps in aluminum blocks of a concentrator. In addition, the procedure involves tens of steps for adding solutions, the repetitive dispensing systems, e.g., Multipette M4 (Eppendorf, Hamburg, Germany) could be used to speed up the solution supply.

## 4 Notes

1. Examine the starch remains in pellets with iodine-potassium iodide solution. If starch is still present after destarch treatments, retreat the samples until the starch is not detected. The remaining starch interferes with the composition analysis. For the starch-free samples, the destarch steps (**steps 9–14** in Subheading 3.1) could be omitted.
2. Sodium azide is a highly hazardous component. Store carefully and strictly follow local waste disposal rules when disposing waste materials containing sodium azide.
3. Weigh samples on a digital scale with the resolution up to 0.01 mg. The accuracy of weighing is important for the assay results.
4. For AIRs, avoid air stream drying to prevent the powder from blowing away.
5. Avoid TFA evaporating by using screw-top tubes and make sure that tubes are sealed completely. Otherwise, the AIRs cannot be fully hydrolyzed. Micro tubes (Polypropylene, 2 ml) from Sarstedt (Nümbrecht, Germany) are used in our lab.
6. Extensive evaporation is necessary to remove the borate as the borate complexes of sugars often affect quantitative recoveries.
7. Avoid acetic anhydride evaporating by sealing the tubes tightly. The loss of acetic anhydride generally causes inaccurate measurement.
8. In our lab, the standard curve is established with the gradients 2, 5, 10, 20, and 30 μg of each monosaccharide and the internal standard 40 μg of inositol. The linear regression coefficient of the standard curve should be larger than 0.99.
9. The identity peak of fucose derivative often elutes around 0.3 min later than that of rhamnose. In addition, the abundance of fucose is generally less than that of rhamnose in wood xylem. Therefore, make sure that the fucose peak is not

mistaken for the rhamnose peak, especially when MSD Chemstation software (Agilent, Santa Clara, CA, USA) is used during auto calculation.

10. Use Microman M250 (Gilson, Villiers-le-Bel, France) or equivalent positive displacement pipettes to handle the viscous and dense liquids, such as concentrated sulfuric acid and anthrone solution.
11. Read adsorptions of the reactions on a spectrophotometer just after cooling to room temperature. Delays of more than 1 h may cause inaccurate measurements.
12. In our lab, the standard curve is established with the gradients 0, 2, 4, 6, 8, 15, and 20 μg of glucose for each assay. Four replicates are conducted. The linear regression coefficient of the standard curve should be larger than 0.99.

## Acknowledgments

This work was supported by the Ministry of Sciences and Technology of China (Grant 2012CB114501) and the National Natural Science Foundation of China (Grants 31530051, 31370310). We thank Dr. Xueqin Song and Prof. Mengzhu Lu (Research Institute of Forestry, Chinese Academy of Forestry) for providing the poplar stems.

### References

1. Watanabe Y, Meents MJ, McDonnell LM, Barkwill S, Sampathkumar A, Cartwright HN, Demura T, Ehrhardt DW, Samuels AL, Mansfield SD (2015) Visualization of cellulose synthases in Arabidopsis secondary cell walls. Science 350:198–203
2. Somerville C, Bauer S, Brinistool G, Facette M, Hamann T, Milne J, Osborne E, Paradez A, Persson S, Raab T, Vorwerk S, Youngs H (2004) Toward a systems approach to understanding plant-cell walls. Science 306:2206–2211
3. Burton RA, Gidley MJ, Fincher GB (2010) Heterogeneity in the chemistry, structure and function of plant cell walls. Nat Chem Biol 6:724–732
4. Scheller HV, Ulvskov P (2010) Hemicelluloses. Annu Rev Plant Biol 61:263–289
5. Scott TA, Melvin EH (1953) Determination of dextran with anthrone. Anal Biochem 28: 1656–1661
6. Dubois M, Gilees KA, Hamilton JK, Rebers PA, Smith F (1956) Colorimetric method for determination of sugars and related substances. Anal Biochem 28:350–356
7. Updegraff DM (1969) Semimicro determination of cellulose in biological materials. Anal Biochem 32:420–424
8. Fry SC (1988) Wall polymer: chemical characterisation. In: Fry SC (ed) The growing plant cell wall: chemical and metabolic analysis. Blackburn, Caldwell, NJ, pp 154–172
9. Albersheim P, Nevins DJ, English PD, Karr A (1967) A method for the analysis of sugars in plant cell-wall polysaccharides by gas-liquid chromatography. Carbohydr Res 5: 340–345
10. York WS, Darvill AG, McNeil M, Stevenson TT, Albersheim P (1986) Isolation and characterization of plant cell walls and cell wall components. Methods Enzymol 118:3–40
11. Foster CE, Martin TM, Pauly M (2010) Comprehensive compositional analysis of plant cell walls (Lignocellulosic biomass): Part II. Carbohydrates. J Vis Exp 37:e1837
12. Pettolino FA, Walsh C, Fincher GB, Bacic A (2012) Determining the polysaccharide composition of plant cell walls. Nat Protoc 7: 1590–1607

13. Sweeley CC, Bentley R, Makita M, Weils W (1963) Gas-liquid chromatography of trimethylsilyl derivatives of sugars and related substances. J Am Chem Soc 85:2497–2505
14. Verhaar LAT, Kuster BFM (1981) Liquid chromtography of sugars on silica-based stationary phases. J Chromatogr 220:313–328
15. Hounsell EF, Davies MJ, Smith KD (1996) Monosaccharide analysis by HPAEC. In: Walker JM (ed) The protein protocols handbook. Humana, Totowa, NJ, pp 635–637
16. Mansfield SD, Kim H, Lu F, Ralph J (2012) Whole plant cell wall characterization using 2D-NMR. Nat Protoc 7:1579–1589

# Chapter 16

# Structural Analysis of Cell Wall Polysaccharides Using PACE

Jennifer C. Mortimer

## Abstract

The plant cell wall is composed of many complex polysaccharides. The composition and structure of the polysaccharides affect various cell properties including cell shape, cell function and cell adhesion. PACE (Polysaccharide Analysis using Carbohydrate gel Electrophoresis) uses a simple, rapid technique to analyze polysaccharide quantity and structure (Goubet et al., Anal Biochem 300:53–68, 2002).

**Key words** Electrophoresis, Polysaccharides, Glycosyl hydrolases, Cell walls, Fluorescent labelling, 8-aminonaphthalene-1,3,6-trisulfonic acid, Xylan

## 1 Introduction

Many techniques for characterizing polysaccharide structure are complicated, requiring expensive equipment and specialized operators, e.g., NMR, MALDI-MS (see also chapters on RAMAN spectroscopy and GC MS in this book). In contrast, PACE provides a rapid and cost-effective method for carbohydrate analysis [1]. While the method here describes xylan analysis, it can also be applied (by use of the appropriate glycosyl hydrolase) to any cell wall polysaccharide.

## 2 Materials

All solutions should be prepared with ultrapure water (18 MΩ at 25 °C) and all chemicals should be analytical grade. All reagents should be prepared and stored at room temperature unless otherwise indicated. Follow local safety guidelines when disposing of waste materials.

Miguel de Lucas and J. Peter Etchells (eds.), *Xylem: Methods and Protocols*, Methods in Molecular Biology, vol. 1544,
DOI 10.1007/978-1-4939-6722-3_16, © Springer Science+Business Media LLC 2017

### 2.1 Equipment

1. SpeedVac equipped with a vacuum pump and cold trap (e.g., Savant concentrator with refrigerated vapor trap, Thermo Scientific). When used, do not heat rotor above 45 °C.
2. Heatblock set at 37, 95 °C.
3. Gel casting apparatus and running apparatus (Hoefer SE660) equipped with a circulating cooling bath (e.g., Hoefer RCB-20) (*see* **Note 1**).
4. High voltage power supply which can provide 1000 V and 250 mA with either constant voltage or constant current (e.g., Owl Hi-voltage TLQ EC1000XL, Thermo Scientific).
5. Hamilton microsyringe (10 μl volume).
6. Pizza cutter.
7. Gel visualization and imaging setup. This should either include a UV-transilluminator equipped with long-wave UV bulbs, a filter set suitable for ANTS (we use a Syngene Chemi XRQ Imaging System) or a laser scanning system (e.g., Typhoon, GE Healthcare).
8. Software for densitometry such as GeneTools (available from Syngene) or the freely available Image J (http://imagej.nih.gov/ij/).

### 2.2 Preparation of Alcohol Insoluble Residue (AIR) from Fresh Plant Material

Alcohol insoluble residue (AIR) is prepared from plant tissue as described in Mortimer et al. (2010) [2].

### 2.3 Glycosyl Hydrolases (GH)

1. GH11 Xylanase was obtained from Megazyme (*see* **Note 2**).
2. Digestion buffer (10×), 1 M ammonium acetate, pH 6.0 (adjust pH with acetic acid). Store at 4 °C, stable for ~3 months (*see* **Note 3**).

### 2.4 Oligosaccharides for Use as Standards

Xylose (X), xylobiose ($X_2$), xylotriose ($X_3$), xylotetraose ($X_4$), xylopentaose ($X_5$), and xylohexaose ($X_6$) are available from Megazyme International (Co. Wicklow, Ireland).

1. Store at −80 °C as 100 mM stocks in water. Working stocks (0.5 ml aliquots) can be diluted to 1 mM in water and stored at −20 °C.
2. (Optional) For quantitative gels, make three different concentration ladders of standards to be run in three different lanes. Standard 1 (S1): mix 1 μl of all six xylo-oligos, S2: mix 2 μl of all six xylo-oligos, S3: mix 5 μl of all six xylo-oligos. For qualitative gels, only one ladder is required (S2 is recommended).
3. Dry standards in the SpeedVac and then fluorescently label as described below.

**2.5 Fluorescent Labelling Components**

1. 2-Picoline-borane (2-PB; Sigma, MO). Make up a 0.2 M stock in dimethyl sulfoxide (DMSO). Since it is *very* hygroscopic, upon opening immediately resuspend the whole amount in DMSO and store in 1 ml aliquots at −20 °C. Stable for at least 1 year.
2. ANTS (8-Aminonaphthalene-1,3,6-trisulfonic acid, disodium salt; Thermo Fisher Scientific, MA), 0.2 M in 17:3 water–acetic acid. Add ANTS to water, warm to 60 °C to dissolve and add the required volume of acetic acid. Store protected from the light at −20 °C. Stable for 4–6 weeks (−80 °C storage may increase stability).
3. DMSO buffer (30 μl acetic acid, 170 μl $H_2O$, 200 μl DMSO)—make fresh each time.
4. 3 M Urea—store at 4 °C for up to 6 months.

**2.6 PACE Gel Components**

1. Forty percent acrylamide–bis-acrylamide (29:1 acrylamide–bis) solution (Sigma, MO) (*see* **Note 4**).
2. 10× resolving buffer (1 M Tris base borate buffer). Prepare ~1.5 M Tris base, and add solid boric acid until the pH is 8.2 (approximately 60 g/L). Dilute solution until Tris final concentration is 1 M.
3. Ammonium persulfate (APS; 10% w/v) in water. Make fresh or use a frozen stock.
4. N,N,N,N′-tetramethyl-ethylenediamine (TEMED).
5. Isopropanol.

## 3 Methods

**3.1 Glycosyl Hydrolase Digestion of AIR**

1. Accurately weigh 20–30 mg AIR into a 15 ml falcon tube (*see* **Note 5**). Resuspend in water to obtain a suspension of 5 mg/ml and vortex/sonicate to ensure even dispersion (a Dounce homogenizer can also be employed). Use a pipette to accurately aliquot the correct amount into individual 1.5 ml microfuge tubes (in this protocol 20 μl, i.e., 100 μg), and then dry in a SpeedVac.
2. Pretreat the dried 100 μg aliquots of Arabidopsis stem AIR with 20 μl of 4 M NaOH for 1 h at room temperature (*see* **Note 6**). Neutralize with 1 M HCl (check pH with pH strips). Add 50 μl of digestion buffer and make the volume up to 500 μl (*see* **Note 7**).
3. Add 3 μl of GH11 (*see* **Note 8**) and incubate overnight at room temperature. Include the following controls (1) AIR no enzyme and (2) enzyme no AIR. Boil the samples at 95 °C in a heat-block for 30 min to deactivate the enzyme. Use microcentrifuge lid locks to prevent loss of sample. Dry samples in a SpeedVac (*see* **Note 9**).

### 3.2 Fluorescent Labelling of Oligosaccharides

1. To each dried sample or standard, add 5 μl of 0.2 M ANTS solution, vortex, and then add 5 μl 2-PB solution and 10 μl of DMSO buffer. Vortex to mix again, and then centrifuge briefly.
2. Incubate at 37 °C for overnight.
3. Dry the samples in the SpeedVac. Resuspend the dried samples in 100 μl 3 M urea, and store at −20 °C for up to 2 weeks (longer at −80 °C) (*see* **Note 10**).

### 3.3 Preparation of ANTS Gel

1. Set up the gel casting apparatus according to the manufacturer's instructions (e.g., http://www.hoeferinc.com/index.php/literature/user-manuals/). To make two resolving gels in an 18 × 24 cm casting set up with 0.75 mm spacers, mix 20.2 ml $H_2O$, 5 ml 10× resolving buffer (1 M Tris-Borate), and 24.6 ml of acrylamide in a 50 ml Falcon tube. Add 200 μl 10 % APS and 20 μl of TEMED, and invert tube a few times to mix. Quickly pour the gel using a large volume pipette (e.g., serological pipette), filling the glass plates to within ~4 cm of the top. If air bubbles get trapped while pouring, tilt the gel or tap the glass to dislodge them.
2. Gently overlay the gel solution with isopropanol. Once the gel has polymerized (~20 min), pour off the isopropanol and rinse with water.
3. For the stacking gel, mix 6.8 ml $H_2O$, 1 ml 10× resolving buffer, and 2.8 ml of acrylamide in a 15 ml Falcon tube. Add 80 μl of APS and 8 μl of TEMED and invert a few times to mix. Rapidly pour stacking gel, and then insert combs, taking care to avoid trapping air bubbles under the comb teeth (*see* **Note 11**).

### 3.4 Separation of Oligosaccharides by PACE

1. Fill running chamber with 1× resolving buffer (0.1 M Tris Borate) and allow it to cool (maintain at 10 °C). This chamber buffer should be changed every 1–2 months, depending on frequency of gel running.
2. Gently remove the comb, and fill the wells with 1× resolving buffer.
3. Load 2 μl of samples or standards per well using a microsyringe. Avoid using the outermost lanes, as these lanes tend to spread on long gels, resulting in the loss of the outermost samples. If preferred, use a mixture of electrophoresis dyes run in the outer lane to visualize the gel progress (e.g., bromophenol blue and thorin1 at 1 mg/ml).
4. Assemble the upper chamber of the gel running apparatus, and place the gel in the cooled lower chamber according to the manufacturer's instructions (e.g., http://www.hoeferinc.com/

index.php/literature/user-manuals/). Gently fill the upper chamber with 1× resolving buffer, without disturbing the samples.

5. Protect the gels from excess direct light, e.g., by wrapping tank with a black plastic garbage bag.
6. Turn on the current, and run for 30 min at 200 V (maintain constant voltage). Then increase voltage to 1000 V for 1 h 40 min (maintain constant voltage) (*see* **Note 12**).

### 3.5 Visualization of PACE Gels

1. Turn on G-box camera (to allow it to cool) and the computer, and open GeneSys software. Clean the transilluminator with water and lint-free tissue (e.g., Kimwipe). Ensure correct filter is in place.
2. View the gel briefly (~80 ms) while still within the glass plates to determine if the dye front is still on the gel. Open gel using wedge tool, and while gel is still on one glass plate, use a pizza cutter to remove both the stacking gel and the bottom of the gel containing any free ANTS dye.
3. Put a small amount of water on the surface of the transilluminator, and, using gloves, place gel directly on the surface of the transilluminator.
4. Take a number of images of varying exposures from short (100 ms) to long (10 s). Ensure at least two do not have any saturating bands (the software can indicate this) (*see* **Note 13**). Do not leave the UV light on between images as this will degrade the fluorescence.
5. Save images as 16 bit Syngene files (.sgd). This is important as it will record information about the image, e.g., exposure time and system setup. Images can be exported as 8 bit tifs for editing in Photoshop, PowerPoint etc.

### 3.6 Interpretation of Results

1. The control lanes are critical to interpreting the gels. Bands that are present in either the enzyme only lane or AIR only lane, as well as the samples should be excluded as "background" (*see* Fig. 1).
2. In order to obtain robust results, it is recommended to repeat PACE analysis on at least three independently grown biological replicates. It is also recommended to perform at least two technical replicates on each biological replicate.
3. The simplest interpretation of data can be through a "fingerprint," e.g., by comparing a wild type plant to a mutant or by comparing the action of two different GHs (*see* Fig. 1).
4. In order to identify bands, either run a standard alongside (e.g., the known products of digestion of glucuronoxylan by GH11 are Xyl, $Xyl_2$ and $MeGlcAXyl_4$, and commercial standards exist) or use further hydrolysis to obtain the oligosaccharide identity,

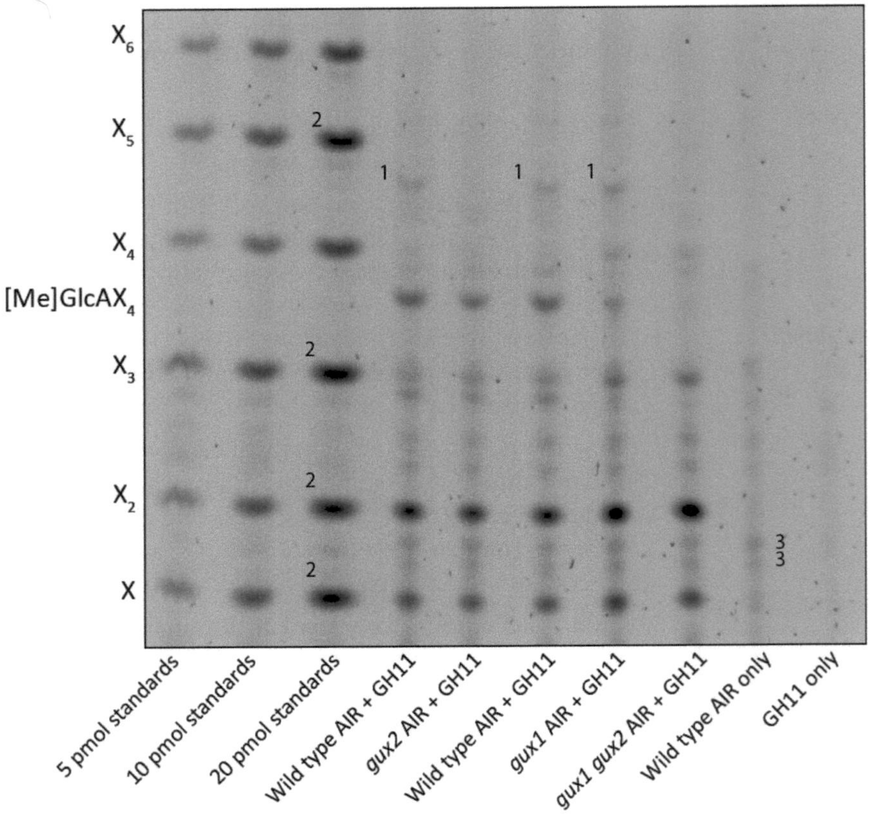


**Fig. 1** Example PACE gel showing a GH11 digestion of wild type (WT) and *gux* mutant Arabidopsis stem as described in [2]. This gel has been specifically chosen as example of a gel with common issues, which are marked up with numbers. (*1*) Issue: more, longer oligosaccharides are released than expected. This indicates incomplete hydrolysis of the xylan by the GH11. Solution: increase quantity of GH added or increase incubation time. (*2*) Issue: bands are saturated. Solution: take images with a shorter exposure. (*3*) Nonspecific bands are identified. Solution: in our experience, it is difficult to track down all sources of nonspecific bands. The controls allow for their exclusion from any analysis

e.g., addition of a xylosidase will hydrolyze $Xyl_2$ to Xyl and $MeGlcAXyl_4$ to $MeGlcAXyl_3$ and Xyl. Addition of a GH115 glucuronidase will hydrolyze $MeGlcAXyl_4$ to $Xyl_4$ etc.

5. Knowledge of all band identities are required in order to quantitate the gel.

### 3.7 Quantitation of PACE Gels

1. After opening the image analysis software of choice, use the three different concentrations of standards to produce an in-gel standard curve. It is recommended to use monosaccharide or disaccharide sugars for this, since commercially available longer oligosaccharides tend to be less pure.
2. *See* Fig. 2 for a description of how to calculate the proportion of xylan branches.

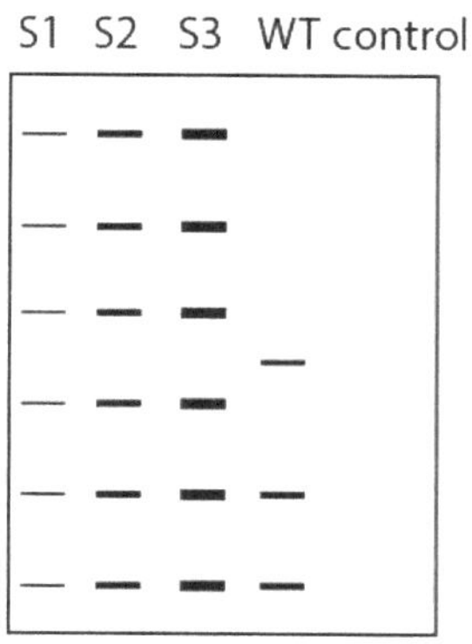


| Peak ID | Peak Area | Quantity (ng/mg AIR) |
|---|---|---|
| Xyl | | |
| $Xyl_2$ | | |
| $[Me]GlcAXyl_4$ | | |
| | | |

$$\text{branching} = \frac{(Xyl) + (Xyl_2 \times 2) + ([Me]GlcAXyl_4 \times 4)}{[Me]GlcAXyl_4}$$

**Fig. 2** Calculating xylan branching frequency from a GH11 xylanase digestion of Arabidopsis stem AIR. Three different concentrations of standards are run within the gel to allow an in gel standard curve to be generated. The Syngene software can then use this to calculate the quantity of oligosaccharides in the sample bands (one fluorescent molecule = one labelled reducing end). Since the identity of all bands are known, the ratio of xylose backbone to [Me]GlcA sidechains can be calculated

## 4 Notes

1. While a standard commercial gel setup could be used (such as that used for SDS-PAGE), the resolution provided by the 18 × 24 cm gels allows for detailed analysis. Thinner gels (0.75 mm spacers) are also recommended.
2. The purity of the GH is critical to the success of the experiment. Prior to usage, it is recommended to test a new GH batch on a range of commercially purified polysaccharide substrates (e.g., homogalacturonan, rhamnogalacturonan I, glucomannan, cellulose, xyloglucan, wheat arabinoxylan) to check for contaminating activities. All GHs used should be heterologously expressed and affinity-tag purified (using fresh affinity resin). Prozomix (www.prozomix.com) make good quality GHs. As an alternative to commercially produced GHs, it is possible to express and purify published enzymes such as the Pichia-expressed GHs described in [3]. The choice of GH depends on the nature of the experiment, but here a GH11 has been used since its digestion products have been well characterized, e.g., [4].
3. Alter the pH conditions to suit the GH used. Ammonium acetate is preferred because it sublimes in the SpeedVac, reducing the salt content of the final sample.

4. Not all acrylamides are made equal, and often cheaper "own-brand" acrylamides don't polymerize consistently in the gel rigs. In my experience, degassing the acrylamide did not alter this.
5. A static eliminator (e.g., AD-1683. A&D Co. Ltd., Saitama, Japan) can reduce the difficulties in weighing this material.
6. This protocol will describe quantities and methods for analyzing the xylan from the basal stem of 6-week-old Arabidopsis plants. The quantity of starting material needed will vary depending on the species, tissue, polysaccharide of interest and GH used. Each reducing end released by the GH will be available for the addition of one fluorescent molecule, so some optimization is required. The severity of the pretreatment step can also be optimized depending on the polysaccharide of interest.
7. If adjusting the pH is proving problematic, add some (~250 μl) water first, as this makes adjusting the pH easier.
8. It is recommended to perform the hydrolysis so that the reaction reaches a stable end point to enable comparison between samples. Therefore when testing a new batch of hydrolase, compare the effects of different protein quantities on the final fingerprint pattern. For a particularly recalcitrant sample, it can be beneficial to add a further 1–2 μl of GH for a second incubation of 1–2 h the following morning.
9. If your samples contain a large amount of particular matter (undigested cell wall; denatured protein etc.), it is recommended to briefly centrifuge the sample and collect the supernatant in a fresh tube. Wash the pellet with a small volume of water, centrifuge again, and combine with the first supernatant. The supernatant can then be dried and derivatized as described.
10. If you intend to produce quantitative data from the gels, then it is recommended that you run the samples within a week of labelling.
11. We prefer to make gels 24 h ahead of running to ensure consistent polymerization and therefore sample separation. Gels can be stored for 1–2 weeks prior to use by covering the top in moist, lint-free tissue, wrapping in a plastic bag and storing at 4 °C. Do not allow the tissue to dry out. Lint-free tissue is key, since most tissues contain optical brighteners which fluoresce in the presence of UV light.
12. If a constant voltage cannot be maintained (especially at higher voltages), this is likely due to a leak in the seal between the upper chamber and the top of the glass plates. Check gasket is properly placed or for damage to glass plates which is preventing a seal being formed. Emergency seals of chips in the glass can be achieved by using a drop of melted 3% (w/v) agarose.

13. If the gel is curling up, or if you are getting background from tissue lint or other dust, it is possible to buy low-fluorescence glass plates and image the gel within the plates. These plates are higher in cost. Ensure the free ANTS has run off the bottom of the gel prior to imaging, as it is extremely bright.

## Acknowledgments

With thanks to Dr. Ramana Pidatala for testing this version of the protocol and to my postdoctoral supervisor, Professor Paul Dupree, whose group developed PACE. This work was part of the DOE Joint BioEnergy Institute (http://www.jbei.org) supported by the U. S. Department of Energy, Office of Science, Office of Biological and Environmental Research, through contract DE-AC02-05CH11231 between Lawrence Berkeley National Laboratory and the U. S. Department of Energy..

### References

1. Goubet F, Jackson P, Deery MJ, Dupree P (2002) Polysaccharide analysis using carbohydrate gel electrophoresis: a method to study plant cell wall polysaccharides and polysaccharide hydrolases. Anal Biochem 300(1):53–68. doi:10.1006/abio.2001.5444
2. Mortimer JC, Miles GP, Brown DM, Zhang Z, Segura MP, Weimar T, Yu X, Seffen KA, Stephens E, Turner SR, Dupree P (2010) Absence of branches from xylan in Arabidopsis gux mutants reveals potential for simplification of lignocellulosic biomass. Proc Natl Acad Sci U S A 107(40): 17409–17414. doi:10.1073/pnas.1005456107
3. Bauer S, Vasu P, Persson S, Mort AJ, Somerville CR (2006) Development and application of a suite of polysaccharide-degrading enzymes for analyzing plant cell walls. Proc Natl Acad Sci U S A 103(30):11417–11422. doi:10.1073/pnas.0604632103
4. Brown DM, Goubet F, Wong VW, Goodacre R, Stephens E, Dupree P, Turner SR (2007) Comparison of five xylan synthesis mutants reveals new insight into the mechanisms of xylan synthesis. Plant J 52(6): 1154–1168. doi:10.1111/j.1365-313X.2007.03307.x

# Chapter 17

# Analysis of Lignin Composition and Distribution Using Fluorescence Laser Confocal Microspectroscopy

**Raphaël Decou, Henrik Serk, Delphine Ménard, and Edouard Pesquet**

## Abstract

Lignin is a polyphenolic polymer specifically accumulating in the cell walls of xylem cells in higher vascular plants. Far from being homogeneous, the lignification of xylem cell walls varies in deposition site, quantity, composition and macromolecular conformation depending on the cell wall compartment, cell type, cell developmental stage and plant species. Here, we describe how confocal microspectroscopy methods using lignin autofluorescence can be used to evaluate the relative lignin amounts, its spatial distribution and composition at the cellular and sub-cellular levels in both isolated cells and histological cross-sections of plant tissues.

**Key words** Lignin, Wood, Xylem, Tracheary elements, Fibers, Vessels, Fluorescence confocal microspectroscopy

## 1 Introduction

Lignin, from the Latin *lignum* 'wood', is a cell wall polyphenolic polymer which varies in deposition site, quantity, composition and macromolecular conformation (1) between the primary and the secondary cell walls, (2) during the differentiation of lignified cells, (3) between cell types and (4) between higher vascular plant species [1]. The chemical diversity of the lignin polymer is due to the oxidative polymerization, in specific cell wall compartments or cell types, of a distinct mixture monolignols, 4-hydroxyphenyl propene monomers which have different hydroxy/methoxylation levels of the benzene ring and different $\gamma$-carbon residues in the aliphatic chain (Fig. 1f) [1].

Particularly abundant in xylem, the hydro-mineral vascular tissue of higher plants [1], lignin can be easily visualized using light microscopy in combination with specific colorigenic methods. These include the Wiesner staining (phloroglucinol in ethanol and HCl) to detect incorporated monolignol aldehydes in purple [2, 5] or the Maüle staining ($KMnO_4/HCl/NaHCO_3$) to distinguish

Miguel de Lucas and J. Peter Etchells (eds.), *Xylem: Methods and Protocols*, Methods in Molecular Biology, vol. 1544,
DOI 10.1007/978-1-4939-6722-3_17, 

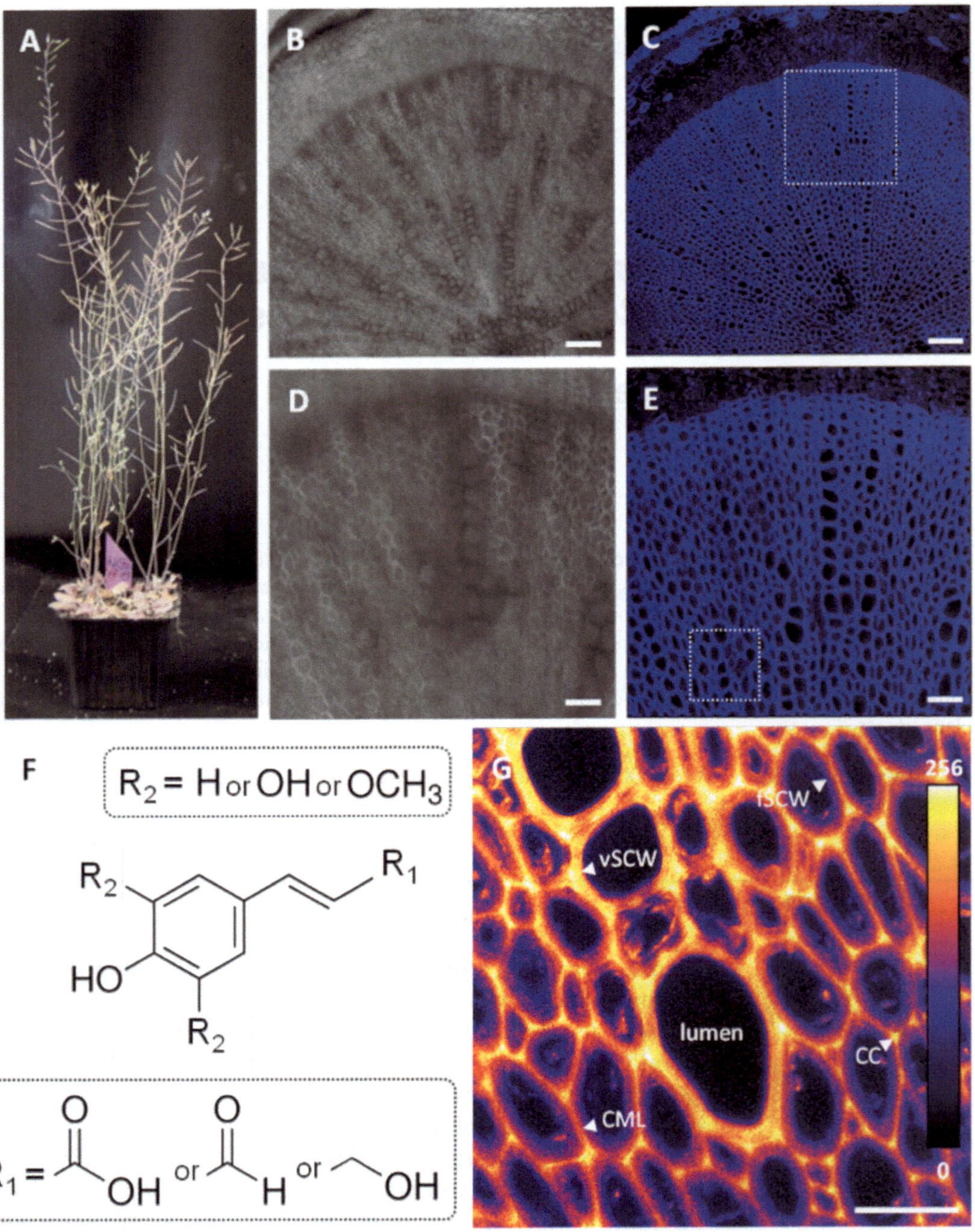


**Fig. 1** Lignin autofluorescence in *Arabidopsis thaliana* stem transversal cross-sections. (**a**) Eight weeks old *Arabidopsis thaliana* plant grown in greenhouse. (**b**) Transversal cross-section of 8-week-old plant hypocotyl observed in brigthfield, bar = 60 μm. (**c**) Transversal cross-section of 8-week-old plant hypocotyl observed in UV fluorescence, bar = 60 μm. (**d**) Close-up (*boxed area* in **c**) of the transversal cross-section observed in brigthfield, bar = 50 μm. (**e**) Close-up (*boxed area* in **c**) of transversal cross-section observed in UV fluorescence, bar = 50 μm. (**f**) Schematic chemical structure of lignin monomers based on 4-hydroxyphenyl propene with different terminal residues of the aliphatic chain (R1) and different substitutions of the aromatic ring (R2). (**g**) Close-up (*boxed area* in **e**) of transversal cross-section observed in UV fluorescence and represented in artificial 8-bit colour-coded scale (from 0 to 256). The different lignified cell wall compartments and cell types are indicated by arrow ends with CML for Compound Middle Lamella, CC for Cell Corner, vSCW for vessel Secondary Cell Wall and fSCW for fiber Secondary Cell Wall, vessel lumen is used as a background control, bar = 50 μm

mono- from bi-methoxylated lignin subunits, stained in orange/brown or red respectively [3, 5]. However, these histological staining methods are limited to the qualitative assessment of lignin deposition and cannot be used to measure the changes in lignin quantity and/or composition at the cellular and sub-cellular levels. More recently, Fourier Transformed Infra-Red (FT-IR) microspectroscopy enabled to determine the relative lignin amount and subunit methoxylation levels in xylem histological sections [5]. However, this method is limited by the use of dehydrated samples, which alters the native spatial organization of tissues [4, 5]. Electron microscopy (EM), using either direct lignin staining with $KMnO4$ [6, 7] or feeding radiolabeled monolignols [8–12], enabled the determination of the spatial distribution of lignin subunits, its amounts and its conformations in specific cell compartments and cell types of xylem. This showed that (a) the quantity of lignin was higher in the middle lamella and cell corners than in the secondary cell walls of tracheary elements (TEs) and fibers [8–12]; (b) the lignin subunit composition was differently enriched between cell wall compartments and cell types: with non-methoxylated in the middle lamella and cell corners, with mono-methoxylated in secondary cell walls of TEs and with bi-methoxylated in secondary cell walls of fibers [8–10]; and (c) the lignin molecules presented different conformations: globular in the middle lamella whereas rod-like in the secondary cell wall [6, 7]. Moreover, EM associated with feeding radiolabeled lignin monomers enabled to estimate the developmental adaptation of xylem cell lignification to gravitational constrains by showing changes in both lignin quantity (reduction) and composition in reaction wood cells [11]. Although EM analyses were extremely efficient in characterizing the quantitative and qualitative changes in lignin between wood cells and compartments, EM sample preparation and image acquisition are labor-intensive and time-consuming, and image quantification relying on non-automated counting and measuring is also a limit.

In contrast to the other cell wall polymersit was shown as early as the sixteenth century that lignin exhibits unique autofluorescent properties [13] (Fig. 1). Indeed, lignin has a broad fluorescent spectrum of multimodal origin constituted of at least four different lignin fluorophores [14, 15] excitable by a range of wavelengths from UV [5, 16–18], to blue [19], green [20] or red [20] light. The most recent lignin autofluorescence microspectroscopy studies were mostly made using UV and blue-light [5, 19]. The fluorophores contributing to lignin autofluorescence depend on the presence of specific chemical residues on the lignin molecule including the type of carbonyl residue (carboxylic, aldehydes or ester) of the aliphatic propene [5, 16] and the numbers of hydroxyl and/or methoxyl residues of the aromatic ring [19] (Fig. 1f). Moreover, the lignin fluorophores intensities are affected by the micro-environments: local pH [19] and other surrounding polymers [21]. Thus the use of fluorescence live-time imaging (FLIM) confirmed different conformational changes of the lignin molecule between middle lamella and secondary cell wall [21]. Hence, lignin autofluorescence microspectroscopy has been used to estimate changes of both composition and spatial distribution in xylem tissues of several species: Arabidopsis [5, 22], zinnia [5], poplar [19], pine [19], spruce [18], eucalyptus [19], acacia [19] and robinia [23]. Lignin autofluorescence microspectroscopy showed similar results to those obtained with EM, including that (a) lignin amounts were differently distributed in xylem—higher amounts in the middle lamella/cell corners than in secondary cell walls and more in tracheary elements than in fibers [5, 22], (b) specific monolignols were differently distributed in xylem [5, 19, 23], and (c) the lignin molecules exhibited different molecular conformation between xylem cell wall compartments [21]. Moreover, alike EM, lignin autofluorescence microspectroscopy showed changes in composition and amounts of lignin in reaction wood cells in response to gravitational constrains [18].

The fluorescent spectra obtained using autofluorescence microspectroscopy can be further processed using multivariate analysis [24] and spectral deconvolution [14, 15]. Altogether, UV and blue-light excited lignin autofluorescence microspectroscopy represents a useful tool to estimate lignin deposition in plant samples by (a) avoiding time-consuming sample preparation, (b) benefiting from confocal microscopy lateral resolution limit of 200–300 nm, and (c) providing spatial quantitative information on lignin relative amounts and composition which can be statistically tested. We herein describe the detail step-by-step methods and possible pit falls associated with the acquisition and signal treatment of UV and blue light-excited lignin autofluorescence microspectroscopy on isolated cells and histological cross-sections of plant tissues.

## 2 Materials

Most of the chemicals were purchase from Sigma-Aldrich (different companies are otherwise specified). All chemicals, contaminated plastics and glassware wastes as well as GMO were eliminated by using the appropriate procedures.

### 2.1 Sample Growth

1. Habituated cell suspension cultures hormonally inducible into TEs

   *Arabidopsis thaliana* habituated cell suspension cultures induced to differentiate into TEs were grown and sampled as described in this issue (Chapter 4) in [25]. Pharmacological treatments with piperonylic acid (PA: inhibitor of lignin monomer biosynthetic enzyme CINNMAMTE-4-HYDROXYLASE

C4H) solubilized in dimethyl sulfoxide (DMSO) were performed as described in [5, 25].

2. Harvesting and storage of stem cross-sections

   Plants are grown in long-day conditions: seeds are incubated in 1 mL of sterile water in a microtube at +4 °C for at least 2 days to accomplish stratification and are planted onto water saturated soil with 16/8 h day/night photoperiod, +22 °C/+18 °C temperature and 70 % relative humidity.

### 2.2 Sample Mounting Medium

To estimate the influence of the mounting medium pH on the emission fluorescence signals (both intensity and spectra), four different mounting media are prepared according to [19]: one unbuffered 50 % glycerol solution prepared in ultrapure water and three 50 % glycerol solutions prepared in 0.01 M phosphate buffer and adjusted with HCl and NaOH to pH 4, 7 or 9 (*see* **Note 1**).

### 2.3 Confocal Microspectroscope

Lignin autofluorescence microspectroscopy requires the use of a spectral confocal microscope. This protocol article describes the use of confocal microscopy as well as the sample preparation, image acquisition and analysis. Here we use a LEICA TCS SP2, equipped with the acousto-optical beam splitter (AOBS), which has been successfully used in several other published studies [5, 22]. This inverted laser scanning confocal microscope has an automated *xyz* stage, which allows to perform three dimensional scanning in both space and emitted light spectra. The two excitation wavelengths chosen to excite lignin autofluorescence are 405 nm (diode UV-laser) or 488 nm (argon laser). Fluorescence emission spectra could thus be obtained by acquiring a 10 nm defined emission window along the light spectra from 410 to 710 nm for a 405 nm laser excitation and 500 to 610 nm for a 488 nm laser excitation.

### 2.4 Image Acquisition and Analysis Software

Image acquisition is made by using the confocal microscope specific software: for example the Leica Confocal Software (LCS) which controls the Leica SP2 confocal microscope. Image treatment, spectral analysis and signal quantification of lignin autofluorescence are performed by using an image processing software, such as the Fiji ImageJ software package (http://imagej.nih.gov/ij/) and its multiple tools including (1) the definition of "region of interest" (ROI) for specific areas and positions in the sample and the “Measure” function of the “Analyze” tab; (2) the compilation of image stacks by using the “Convert images to stack” function of the “Image” tab; (3) the “Time Series Analyzer V3” plug-in for volumetric and spectral measurements which manages the acquisition of multiple ROIs for every image on a stack; and (4) the “Adjust” function in the “Image” tab to perform projection (“sum of stack” on “z project”) and the colour coding using the 8-bit “Fire” colour scale using the “Look-up tables” of the “Image” tab.

### 2.5 Graph and Statistical Analysis Software

Graphs and statistical analysis are produced using Microsoft Excel by importing the spectral and volumetric data measured using ImageJ onto spreadsheets.

## 3 Methods

All experiments are carried out at room temperature while samples are stored at +4 °C prior to observation and analysis.

### 3.1 Harvesting and Sample Storing

1. Harvesting and storage of cell cultures

   Under a sterile laminar flow hood, 25–30 μL aliquots of *Arabidopsis thaliana* habituated cell suspension cultures induced to differentiate into TEs [25] and treated or not with PA are sampled by pipetting from larger suspension cultures in 250 mL Erlenmeyer flasks. Sampled cells are stored in a microtube at +4 °C prior to observation.

2. Harvesting and storage of stem cross-sections

   Eight weeks after germination (Fig. 1a), 0.1–0.3 mm thick transversal sections of *Arabidopsis thaliana* wild-type or mutant plants are hand-sectioned with a razor blade at the basal part of the floral stem (Fig. 1b–e). Newly made transversal sections are placed directly in 2 mL microtubes containing 1 mL of 70 % ethanol solution and stored at −20 °C for at least 1 week before use. This step is essential to remove cell protoplasts and other intracellular autofluorescent compounds such as chlorophyll.

### 3.2 Sample Preparation for Confocal Microspectroscopy

1. Sample mounting for observations (cell cultures and cross-sections)

   Isolated cells obtained from cell suspension cultures are directly mounted between microscope glass slide and cover slip. For the stem cross-sections, the 70 % ethanol storage solution is removed and replaced by 1 mL ultrapure water. The washed sections are then incubated in the un-buffered or buffered mounting media, depending on the sample, for at least 15 min before confocal observation as initially described in [19]. To avoid damaging the cross-section, medium and samples are transferred from the microtube to the microscope slide using a cut 1 mL pipette tip (*see* **Note 2**). Once the sample is mounted and prior to clamping the microscopy slide in the brackets of the microscope stage, a drop of distilled water or immersion liquid is deposited on the immersion objective (HCX PL APO 63.0 × 1.20 W BD UV).

2. Calibration of fluorescence emission

   The calibration of the fluorescence is performed on the control samples (i.e. wild-type or non-treated cell cultures). To do so, the *z* depth is first adjusted to focus on the sample (LCS: using the "Visual" mode). Subsequently different parameters such as the laser excitation, the laser intensity and the emission wavelength collection window are defined (LCS: using the "Beam Path Settings" window). These parameters include: (1) the beginning and ending wavelength of the "PMT" (photomultiplier tube) collecting the fluorescent emitted light (*see* **Notes 3** and **4**); (2) the transmitted light (LCS: using "PMT Trans") for brightfield image acquisition (*see* **Note 4**); (3) the laser intensity (expressed in %); (4) the focus of the exciting laser beam (LCS: using "Beam expander"); (5) the speed to measure photons in the collection window (LCS: using "Scan speed"); (6) the confocality of the image acquired using "Pinhole" (expressed in airy units); (7) the averaging using line or frame averages (LCS: "Li. A." and "Aver" buttons); (8) the position in the sample using the *xyz* volumetric mode to define top and bottom of the sample or the *xyλ* spectral mode to define the beginning and the end of the emission collection window. The digital calibration to an 8-bit scale of the intensity of the emitted light for one set laser intensity is adjusted (LCS: "Continuous" mode) together with the colour-coded image scale (LCS: "Q-LUT" mode) and the Smart Offset potentiometer in order to set the background to 0 (shown in green) and the highest emission to 256 (shown in blue).

3. Image format and acquisition parameters

   The image formats are set to 512x512 pixels image in 8-bits. The image acquisitions are performed with a scan speed (LCS: "Speed" button) of 400 Hz for both single scan (LCS: 2D *xy* scan—"Single scan" button) or stack series (LCS: 3D *xyz* or spectral *xyλ* images—"Series" button), chosen according to the analyses performed.

### 3.3 Confocal Microscopy Volumetric Imaging

1. Parameters for volumetric acquisition

   Fluorescence confocal microscopy enables to obtain fluorescence data and to optically section the sample at different depths to obtain a three-dimensional image (referred to *xyz*). Depending on the desired excitation wavelength, either the 405 nm diode UV-laser or the 488 nm argon lasers are used (LCS: using the "Beam path settings" window). Subsequently, the corresponding PMT—PMT1 for 405 nm laser excitation or PMT2 for 488 nm laser excitation—are activated and the band range of the emission spectra are set to 409–469 nm or to 509–568 nm, respectively. The "*xyz*" scan mode is selected (LCS: "Mode" button): both the line average and the frame average are set to 2 (LCS: "Li. A." and "Aver" buttons respectively) in order to improve the signal-to-noise ratio. The volumetric acquisition for isolated TEs, cells and plant transversal cross-sections are obtained similarly-except for the image averaging with line and frame averages set to 1 and 6 (single scan acquisition) or 1 and 3 (serial images acquisition) respectively (Fig. 2). Using continuous image acquisition (LCS: "Continuous"

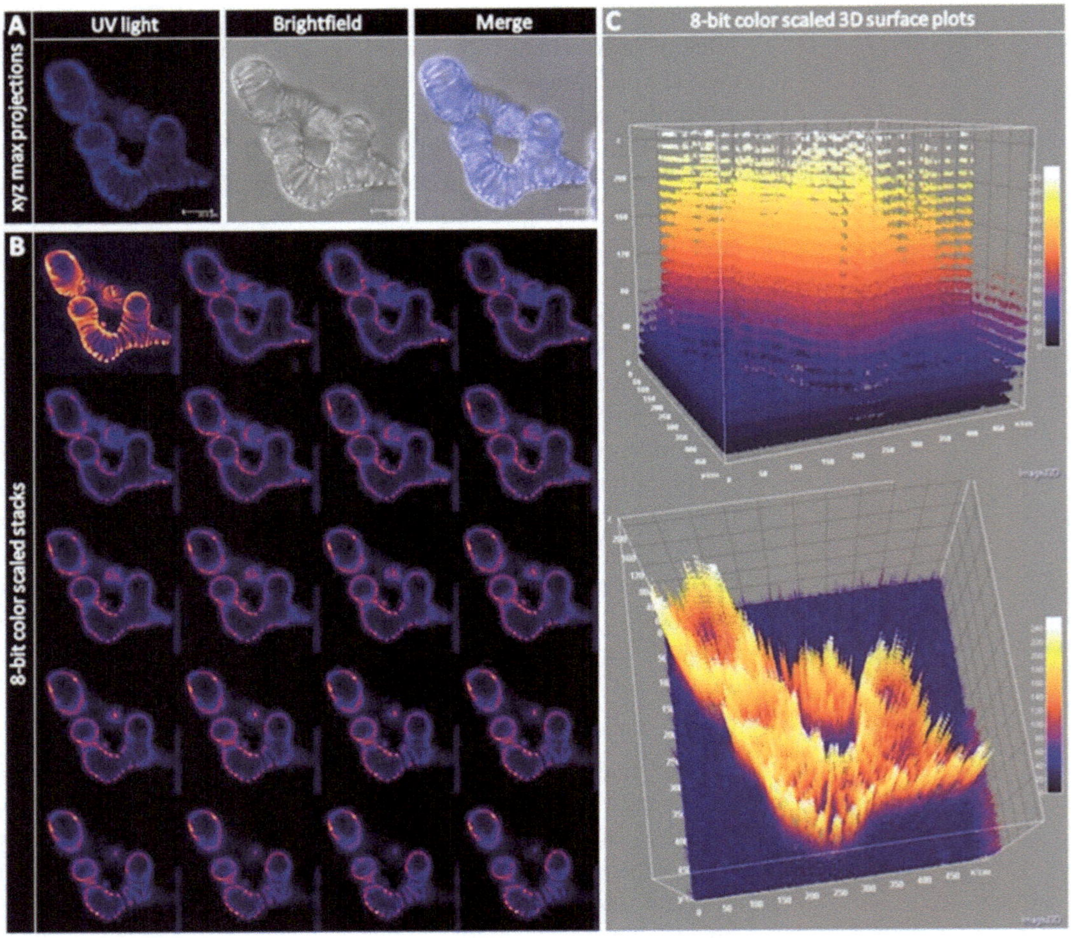


**Fig. 2** Observations of lignin autofluorescence by UV-confocal microscopy of *Arabidopsis thaliana* TEs. (**a**) Maximal projection of *xyz* image stacks of TEs observed in UV fluorescence, brightfield and the merge of both, bars = 10 μm. (**b**) Volumetric *xyz* image serie through TEs observed in UV fluorescence and represented in artificial 8-bit colour scale (from 0 to 256). The first image (in the top left-hand corner) is a merged image of the 19 other non-volumetric xy stacks. (**c**) Variation of 8-bit fluorescence intensity in *xyz* image series through TEs observed in UV fluorescence (8-bit colour scaled 3D surface plots)

button), the upper and lower limits of the *z*-range are selected (LCS: small "Series" button) based on the visual boundaries of the fluorescent signal in both directions (a 10–15 μm depth is generally chosen). The general amount of sections is set to 20 (LCS: "Sect" button) in order to obtain a step size of about 0.5–0.75 μm between each images during the serial image acquisition (LCS: large "Series" button activated).

2. Calibration of fluorescence

   The laser intensity is set to 100 % (LCS: "Beam path settings" window) for all excitation wavelengths. For the 405 nm excitation, the PMT is adjusted to a maximum of 888 or 1000 V, depending on the type of analyzed sample. The focus of the exciting laser beam (LCS: "Beam expander" button) is set to 6 and the "Pinhole" adjusted to 1 airy unit. For the 488 nm excitation, the PMT voltage is reduced to a maximum of 600 V (LCS: "Signal" button). Due to a lower emitted fluorescent signal, the exciting laser beam focus (LCS: "Beam expander") is set to 3 and the "Pinhole" to 2 airy units. The colour-coded image scale is adjusted (LCS: "Q-LUT" function combined with the "Smart Offset") to ensure that the 8-bit background signal equals 0 and that the most intense fluorescence of the sample is below saturation (below 256).

3. Image analysis and quantification
   (a) Analysis method

   The 8-bit grey scale values are obtained using the image processing software ImageJ. The analysis of the fluorescence intensity in different areas of the plant section volumetric scans is obtained by projecting the image series onto one image using the "sum of stack" feature on "*z* project" for each stacks. Whereas in intact isolated cells produced in cell suspension culture, the fluorescence signal is obtained directly from single 2D scan images (Fig. 3).

   (b) Measurements

   Circular ROIs of 200–1000 nm diameter, corresponding to 20–50 pixels (*see* **Note 5**), are set at different positions to measure lignin fluorescence in the different cell wall compartments. These include: the lumen, the vessel secondary cell wall (vSCW), the fiber secondary cell wall (fSCW), the cell corner (CC) and the compound middle lamella (CML) in stem cross-sections (Fig. 1g). For isolated cells, circular ROIs of 500–2000 nm diameter (150–500 pixels, depending on the magnification) delimit the outside of the cell, the primary cell wall (PCW) and the secondary cell wall (SCW) (Fig. 3) (*see* **Note 6**). The background fluorescence is measured using the lumen for transversal sections or the areas outside of the isolated cells. A minimum of 5 ROI is set for each cell wall compartment and cell type analyzed using the ROI manager of the "Time Series Analyzer V3" plug-in. The fluorescence intensity is obtained for each position using the "Get Average" button. All the values are then exported to Microsoft Excel spreadsheets using copy/paste.

   (c) Normalization and graphing

   A standard normalization is performed for each individual plant section to express lignin autofluorescence as a percentage of the total fluorescence measured. To do so, the sum of all measurements is calculated in 8-bit grey

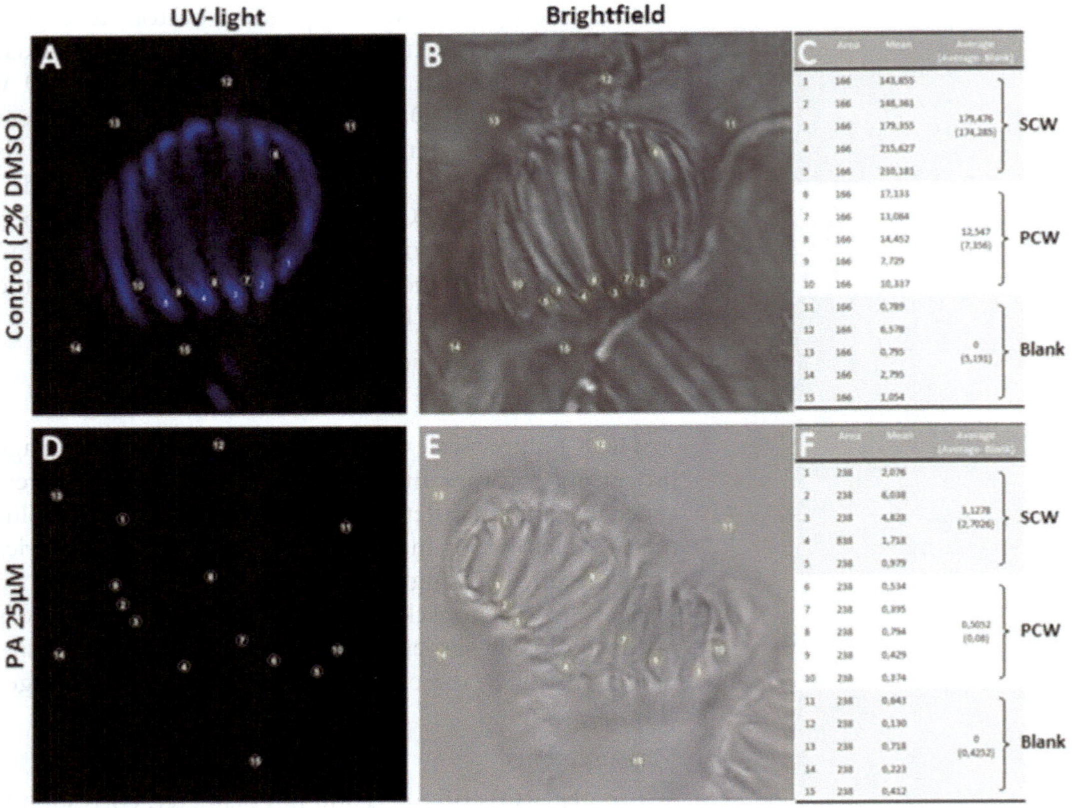


**Fig. 3** Quantification of lignin autofluorescence in *Arabidopsis thaliana* TEs treated or not with the lignin monomer biosynthesis inhibitor PA. (**a**) *xy* single image of TE treated with 2 % DMSO observed in UV fluorescence, the numbers indicate the position of the defined measurement areas outside the cell and in the primary and secondary cell walls (from 1–15). (**b**) *xy* single image of TE treated with 2 % DMSO observed in brightfield, the numbers indicate as above (from 1–15). (**c**) Measurements of the UV fluorescence intensity in the selected positions in (**a**): the numbers 1–5 represent the secondary cell wall (SCW), 6–10 the primary cell wall (PCW) and 11–15 the outside of the cell. (**d**) *xy* single image of TE treated with 25 µM piperonylic acid (PA) observed in UV fluorescence, the numbers indicate the position of the defined measurement areas outside the cell and in the primary and secondary cell walls (from 1–15). (**e**) *xy* single image of TE treated with 25 µM piperonylic acid (PA) observed in brightfield, the numbers indicate as above (from 1–15). (**f**) Measurements of the UV fluorescence intensity in the selected positions in (**d**): the numbers 1–5 represent the secondary cell wall (SCW), 6–10 the primary cell wall (PCW) and 11–15 the outside of the cell

scale values and each measurement is divided by the sum of measured areas (resulting in a normalization based on the measured area). To obtain a normalized distribution of the grey scale values, the sum of all the different cellular positions is calculated and each position is divided by this sum, resulting in a relative percentage of the lignin fluorescence intensity for each cellular/subcellular position. For isolated cells, the sum of the grey scale values is divided by the number of selected positions, resulting in an average based on the number of areas analyzed. For each cellular compartment, the sum of all these averages from all the replicates is divided by the total number of TEs analyzed in all the replicates. This normalization resulted in a TE cell wall distribution of the lignin fluorescence intensity according to all the merged replicates.

### 3.4 Confocal Microspectroscopy Spectral Imaging

The broad autofluorescence of lignin results from the contribution of multiple lignin fluorophores exhibiting specific accumulations, localizations, sensitivity to the microenvironment, and fluorescence quantum yields. These lignin fluorophores directly depend on the chemical structure of the lignin molecules such as the carbonyl residue type (carbonyl vs alcohol) of the subunit aliphatic propene chain [5, 16]. The acquisition of the lignin fluorescence spectra therefore provide key compositional information on the polymer. For example, it detects the presence of syringyl γ-p-hydroxybenzoates residues—like between pine and poplar [19]—or the increase of carbonyl residues in the *Arabidopsis thaliana* loss-of-function mutant in *CINNAMOYL-COA REDUCTASE ccr1-3* [5] (Fig. 4).

1. Parameters for spectral acquisition

   To perform the spectral measurement of lignin autofluorescence, the *xyλ* mode is selected (LCS: "Mode" button) and the line average and frame average (LCS: "Li. A." and "Aver" button) are both set to 0. The beginning and the end of the emitted light collection windows are set to 410–710 nm for 405 nm laser excitation or to 500–610 nm for the 488 nm laser excitation (LCS: "λ Beg." and "λ End" buttons). The step size in-between the limits of the emitted light collection window is set to 30 (LCS: "λ Step" button) for both 405 and 488 nm laser excitations, thus leading to a spectral width of 10 and 4 nm respectively.

2. Calibration of fluorescence

   Calibration of the fluorescence is adjusted as described for the volumetric acquisition in **step 2** of Subheading 3.2. although the PMTs are generally set to their maximum voltage due to the small width of the collection window.

3. Spectra and image analyses

   (a) Spectral measurements

      Using ImageJ "oval" selection button and the "Time Series Analyzer V3" plug-in, circular ROIs with a diameter ranging from 200 nm to 2 μm are marked for different cellular or subcellular positions of TEs from cell cultures (Fig. 3) or for different positions (vSCW, fSCW, CC, CML and vessel lumen as a fluorescence background control) throughout the cross-section (Fig. 4) (*see* **Note 7**). Spectra are then measured using the "Get Average" button of the "Time Series Analyzer V3" window.

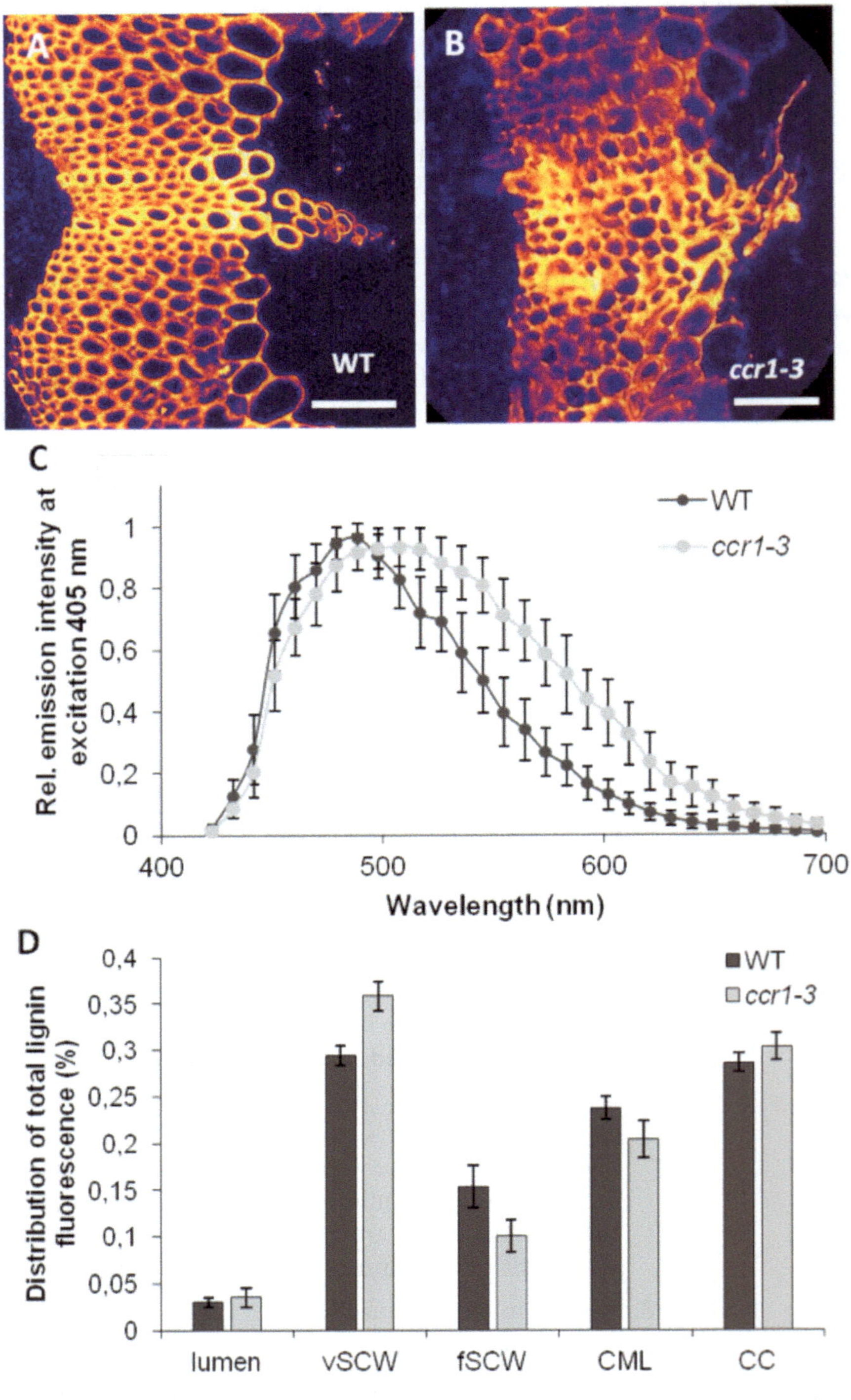


**Fig. 4** Lignin autofluorescence spectra and spatial distribution in *Arabidopsis thaliana* wild-type and *ccr1-3* mutant stem cross-sections. (**a**) Stem transversal cross-section of 8-week-old wild-type (WT) observed in UV fluorescence represented in 8-bit artificial colour scale (from 0 to 256), bar = 50 μm. (**b**) Stem transversal cross-section of 8-week-old *ccr1-3* mutant observed in UV fluorescence represented in 8-bit artificial colour scale (from 0 to 256), bar = 50 μm. (**c**) Lignin UV fluorescence spectra of WT and *ccr1-3* mutant plants, note that the spectrum of the *ccr1-3* mutant is shifted by 38 nm from the WT spectrum. Error bars indicate ± SD (n = 5 replicates). (**d**) Spatial distribution of the lignin UV fluorescence in stem cross sections of WT and *ccr1-3* plants for the different cell wall compartments and cell types of xylem including lumen, vessel secondary cell wall (vSCW), fiber secondary cell wall (fSCW), compound middle lamella (CML) and cell corner (CC). Error bars indicate ± SD (n = 5 replicates)

(b) Normalization and graphing

The spectral data are then exported to Microsoft Excel and are range-normalized by dividing each value by the maximum value of the spectrum thus restricting the range between 0 and 1. The average spectrum and its standard deviation are subsequently obtained from each wavelength of the different range-normalized spectra. The statistical analysis, using a Student *t*-test, is then performed for specific wavelengths of the range-normalized spectra of different plants, such as between wild-type *Arabidopsis thaliana* and the loss-of-function mutant *ccr1-3* (Fig. 4) [5].

## 4 Notes

1. Note that the influence of the mounting medium on the fluorescence intensity and spectrum depend on the species used. To estimate the influence of the microenvironment on lignin autofluorescence, test a series of 50 % glycerol buffered in 0.01 M phosphate at pH 4, 7 and 9.
2. To avoid the mechanical destruction of the suspended cells or cross-sections by the pipette tip, cut the end of the tip with scissors before performing the pipetting steps.
3. In order to avoid reflection, choose an emission window which is at least 8 nm above the laser excitation wavelength.
4. The gain values of each PMT is expressed in volts (V).
5. To get optimal results, measure 5–10 positions for each cell type or cellular position. Measuring tools are saved and activated using the ROI manager function of ImageJ.
6. For each TE, five identical circular areas were selected in each cell compartment (SCW and PCW) and the blank. Five to ten TEs/cell culture were measured with the ROI manager function of ImageJ.
7. For optimal results, measure at least three different positions for each cellular or subcellular position.

## Acknowledgment

This work was supported by Vetenskapsrådet VR (grant 2010-4620 to E.P.), the BioImprove FORMAS (to E.P.), the Berzelii Centre for Forest Biotechnology (to D.M. and E.P.), by the Kempe Foundation (Gunnar Öquist Fellowship to E.P.), by the Carl Tryggers Foundation and by the Stiftelsen för Strategisk Forskning (ValueTree) to E.P. We also thank the Swedish research council Vetenskapsrådet and the Swedish Governmental Agency for Innovation Systems Vinnova (UPSC Berzelii Centre) and Bio4Energy (a strategic research environment appointed by the Swedish government).

## References


1. Barros J, Serk H, Granlund I, Pesquet E (2015) The cell biology of lignification in higher plants. Ann Bot 115:1053–1074
2. Pomar F, Merino F, Barceló AR (2002) O-4-Linked coniferyl and sinapyl aldehydes in lignifying cell walls are the main targets of the Wiesner phloroglucinol-HCl reaction. Protoplasma 220:17–28
3. Crocker EC (1921) An experimental study of the significance of "lignin" color reactions. Ind Eng Chem 13:625–627
4. Gorzsás A, Stenlund H, Persson P, Trygg J, Sundberg B (2011) Cell specific chemotyping and multivariate imaging by combined FT-IR microspectroscopy and OPLS analysis reveals the chemical landscape of secondary xylem. Plant J 66:903–914
5. Pesquet E, Zhang B, Gorzsás A, Puhakainen T, Serk H, Escamez S, Barbier O, Gerber L, Courtois-Moreau C, Alatalo E, Paulin L, Kangasjärvi J, Sundberg B, Goffner D, Tuominen H (2013) Non-cell autonomous post-mortem lignification of tracheary elements in Zinnia elegans. Plant Cell 25:1314–1328
6. Donaldson L (1992) Lignin distribution during latewood formation in Pinus radiata D. Don. IAWA Bull 13:381–387
7. Fromm J, Rockel B, Lautner S, Windeisen E, Wanner G (2003) Lignin distribution in wood cell walls determined by TEM and backscattered SEM techniques. J Struct Biol 143:77–84
8. Terashima N, Fukushima K, Tsuchiya S, Takabe K (1986) Heterogeneity in formation of lignin: VII. An autoradiographic study on the formation of guaiacyl and syringyl lignin in poplar. J Wood Chem Technol 6:495–504
9. Terashima N, Fukushima K (1988) Heterogeneity in formation of lignin: XI. An autoradiographic study of the heterogeneous formation and structure of pine lignin. Wood Sci Technol 22:259–270
10. Terashima N, Fukushima K, Sano Y, Takabe K (1988) Heterogeneity in formation of lignin: X. Visualization of lignification process in differentiating xylem of pine by microautoradiography. Holzforschung 42:347–350
11. Fukushima K, Terashima N (1991) Heterogeneity in formation of lignin: Part XV. Formation and structure of lignin in compression wood of Pinus thunbergii studied by microautoradiography. Wood Sci Technol 25:371–381
12. Terashima N, Yoshida M, Hafrén J, Fukushima K, Westermark U (2012) Proposed supramolecular structure of lignin in softwood tracheid compound middle lamella regions. Holzforschung 66:907–915
13. Monardes N (1565) Historia medicinal de las cosas que se traen de nuestras Indias Occidentales que sirven en medicina. Sevilla
14. Djikanović D, Kalauzi A, Radotić K, Jeremić M (2007) Deconvolution of lignin fluorescent spectra: a contribution to the comparative structural studies of lignins. Russ J Phys Chem 81:1425–1428
15. Donaldson L, Radotić K, Kalauzi A, Djikanović D, Jeremić M (2010) Quantification of compression wood severity in tracheids of Pinus radiata D. Don using confocal fluorescence imaging and spectral deconvolution. J Struct Biol 169:106–115
16. Albinsson B, Li S, Lundquist K, Stomberg R (1999) The origin of lignin fluorescence. J Mol Struct 50:19–27
17. Donaldson LA, Singh AP, Yoshinaga A, Takabe K (1999) Lignin distribution in mild compression wood of Pinus radiata. Can J Bot 77:41–50
18. Donaldson LA, Radotic K (2013) Fluorescence lifetime imaging of lignin autofluorescence in normal and compression wood. J Microsc 251:178–187
19. Donaldson LA (2013) Softwood and hardwood lignin fluorescence spectra of wood cell walls in different mounting media. IAWA J 34:3–19
20. Pesquet E, Tuominen H (2011) Ethylene stimulates tracheary element differentiation in Zinnia cell cultures. New Phytol 190:138–149
21. Zeng Y, Zhao S, Wei H, Tucker MP, Himmel ME, Mosier NS, Meilan R, Ding SY (2015) In situ micro-spectroscopic investigation of lignin in poplar cell walls pretreated by maleic acid. Biotechnol Biofuels 8:126–138

22. Serk H, Gorzsás A, Tuominen H, Pesquet E (2015) Cooperative lignification of xylem tracheary elements. Plant Signal Behav 10:e1003753
23. Nakaba S, Yamagishi Y, Sano Y, Funada R (2012) Temporally and spatially controlled death of parenchyma cells is involved in heartwood formation in pith regions of branches of Robinia pseudoacacia var. inermis. J Wood Sci 58:69–76
24. Radotić K, Kalauzi A, Djikanović D, Jeremić M, Leblanc RM, Cerović ZG (2006) Component analysis of the fluorescence spectra of a lignin model compound. J Photochem Photobiol 83:1–10
25. Ménard D, Serk H, Decou R, Pesquet E (2016) Establishment and utilization of habituated cell suspension cultures for hormone-inducible xylogenesis. In: Xylem. Springer, New York

# Chapter 18

# Topochemical Analysis of Cell Wall Components by TOF-SIMS

Dan Aoki and Kazuhiko Fukushima

## Abstract

Time-of-flight secondary ion mass spectrometry (TOF-SIMS) is a recently developing analytical tool and a type of imaging mass spectrometry. TOF-SIMS provides mass spectral information with a lateral resolution on the order of submicrons, with widespread applicability. Sometimes, it is described as a surface analysis method without the requirement for sample pretreatment; however, several points need to be taken into account for the complete utilization of the capabilities of TOF-SIMS. In this chapter, we introduce methods for TOF-SIMS sample treatments, as well as basic knowledge of wood samples TOF-SIMS spectral and image data analysis.

**Key words** TOF-SIMS, Cellulose, Lignin, Extractives, Inorganic metals, Chemical mapping

## 1 Introduction

Time-of-flight secondary ion mass spectrometry (TOF-SIMS) is one of the most sensitive surface analysis techniques as it provides mass spectral information with a lateral resolution on the order of submicrons [1]. A primary ion beam (such as $Ga^+$, $Cs^+$, $Au_n^+$, $Bi_n^+$, $Ar_n^+$, and $O^-$) is applied to the sample surface. A mass spectrometer detects the secondary ions sputtered from the sample surface that are result of the primary ion bombardment. The schematic illustration of chemical mapping by TOF-SIMS is shown in Fig. 1. Inorganic and organic ions are detected in the positive and negative ion modes. In the TOF-SIMS spectra of wood materials, small molecules and inorganic metals are detected, such as $K^+$ with an $m/z$ of 38.96, where $m/z$ is the mass to charge ratio.

In this chapter, we introduce methods for TOF-SIMS sample treatments. In addition, the basic knowledge of TOF-SIMS spectral and image data analysis for wood samples is also briefly described. It is hoped that readers will expand their analytical alternatives after this chapter.

Miguel de Lucas and J. Peter Etchells (eds.), *Xylem: Methods and Protocols*, Methods in Molecular Biology, vol. 1544, DOI 10.1007/978-1-4939-6722-3_18, © Springer Science+Business Media LLC 2017

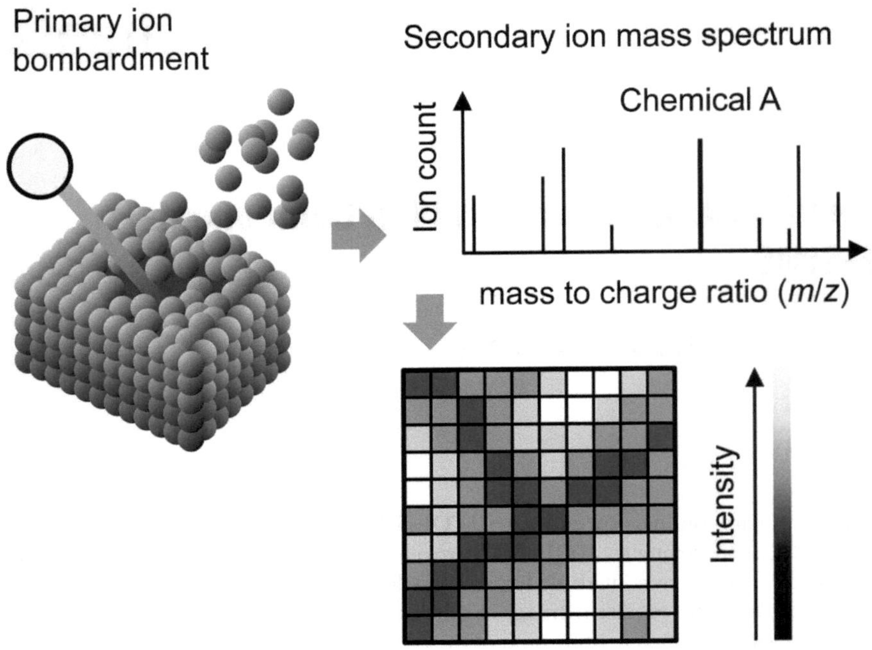


**Fig. 1** Schematic illustration of TOF-SIMS measurement visualizing the distribution of chemical A

## 2 Materials

### *2.1 Sample Manipulation*

In TOF-SIMS measurements, the distance below the surface is estimated as a few nanometers; therefore, the quality of data significantly depends on surface conditions. i.e., if a 10 mm cubic wood block is applied for measurements, the 10 nm surface might be contaminated by some chemicals. This pollutant concentration represents only 1 ppm for the whole block, but it is 100% at the measurement surface and, consequently, the resultant spectrum is solely attributed to the pollutant.

One of the most widely used organic pollutants is poly (dimethylsiloxane) (PDMS); it exhibits high chemical stability and is used widely as a mold lubricant, plasticizer, surface modifier, antifoaming agent and as a cleaner additive. PDMS can easily pass from the surrounding materials to the sample surface. It is important to remove PDMS from the sample, as the secondary ion yield of PDMS is very high. To remove PDMS, washes with *n*-hexane are effective; however, complete PDMS removal is difficult. Furthermore, solvents could also wash some of the sample original chemicals.

For TOF-SIMS measurements, we recommend the use of quantitative filter papers, polypropylene-coated papers, polypropylene tools, aluminum foil (only for dried sample), and well-washed glass tools for sample preservation. New glass tools should be washed with *n*-hexane before use because PDMS is present on their surfaces. In addition, all experimental items such as disposable pipettes, tweezers, and spatulas should be washed to remove other types of pollutants such as inorganic metals, salts, and oils.

Surface flatness is also an important factor for optimal measurements. Typically, the incidence angle of the primary ion beam is tilted for maximizing the secondary ion yield (Fig. 1). If the sample surface is uneven, the ion beam would not irradiate some area and the image will contain a dark spot. The uneven surface may also result in low mass resolution.

Before TOF-SIMS measurements, a convenient, general method for preparing surfaces might involve cutting using a microtome or a razor blade. Various preparation methods for traditional microscopic observations are available for preparing flat, clean surfaces; however, the chemical treatments associated with them (such as extraction and resin embedding) might potentially change the sample chemical composition. Every preparation method should be designed by taking into account the target chemicals of the measurements. Example procedures for plant and standard samples are described below.

1. Wood samples for cell wall polymer components
   (a) Cut a wood sample into a small block using a saw, a chisel, pruning shears, razors, or any other wood working tools (*see* **Note 1**).
   (b) Cut a thin section from the block as a measurement surface. Immersing the block in water might help in easily cutting the block. The appropriate thickness of the section depends on the measurement machine (*see* **Note 2**).
   (c) Wash the section by solvents if needed. Washing by organic solvents, such as acetone, *n*-hexane, and methanol, can remove soluble extractives from the sample and increase ion intensities of the insoluble polymer components or cellulose, hemicellulose, and lignin.
2. Wood samples for low molecular extractives
   (a) Freeze–drying might be better for the drying of the harvested plant samples. The so-called heartwood extractives are mainly insoluble in water, but the distribution can change while the drying period [2, 3].
   (b) Cutting procedures are nearly the same as those described in **item 1**.
3. Wood samples for water-soluble chemicals
   (a) Frozen samples and a cryo-TOF-SIMS apparatus are needed for visualizing water-soluble chemicals, and several research groups have reported special procedures [3–6]. We suggest the simplified method described below. Standard aqueous solutions are measured by the same procedure.

(b) Rapid freezing techniques should be better to maintain living structures.

(c) Cut the frozen-hydrated sample using a cryo-microtome and transfer the sample to the TOF-SIMS apparatus equipped with a cryostage, avoiding any melting or frosting of the sample surface.

4. Powder samples.
   Because of the high-vacuum conditions and the irradiation impact of the primary ion beam for TOF-SIMS measurements, it is better to fix the powder sample on the base material, such as a soft metal plate (Au or In) and several adhesive materials (*see* **Note 3**). The base material and possible impurities should be measured in advance for distinguishing the secondary ions of the target sample (*see* **Note 4**).

5. Others.

   Solvent-soluble samples can be casted on solid base materials and measured as a flat surface. Silicon wafer and glass plates are available. This thin-layer preparation method is also applicable to nonvolatile liquid samples, such as unsaturated fatty acids. The adsorption on a filter paper is a possible approach for nonvolatile chemicals. For volatile chemicals, a cryo-sample stage should be worthwhile for further study.

## 3 Methods

### 3.1 Measurements

Typically, a TOF-SIMS apparatus has two measurement modes for image and spectrum analyses. To improve mass resolution, the primary ion beam is bunched to make the beam length short. As a result of the bunching, the lateral resolution decreases slightly in the spectrum mode by the widened primary ion beam. The following text does not distinguish these modes.

### 3.2 Spectrum Analysis

In TOF-SIMS measurements, the maximum of a detectable $m/z$ value might be a several thousands [1]. It is nearly impossible to detect a macromolecule as it is, and several fragment ions in the mass spectrum will be observed. Polymers having repeating units tend to be cleaved at the interunit linkage. In this section, secondary ions of plant cell wall components are briefly introduced. Their detailed fragmentation behavior is summarized elsewhere [7].

The repeating unit of cellulose is anhydroglucose [$C_6H_{10}O_5$], and the monomeric and oligomeric fragment ions such as $[C_6H_{10}O_5]_n[H]^+$ ($m/z$ 163, 325, 487, 649, etc.) are detected. After the dehydration of the monomeric unit, fragment ions such as $[C_6H_8O_4][H]^+$ ($m/z$ 145) and $[C_6H_6O_3][H]^+$ ($m/z$ 127) are also detected. As a result of further fragmentation, several smaller

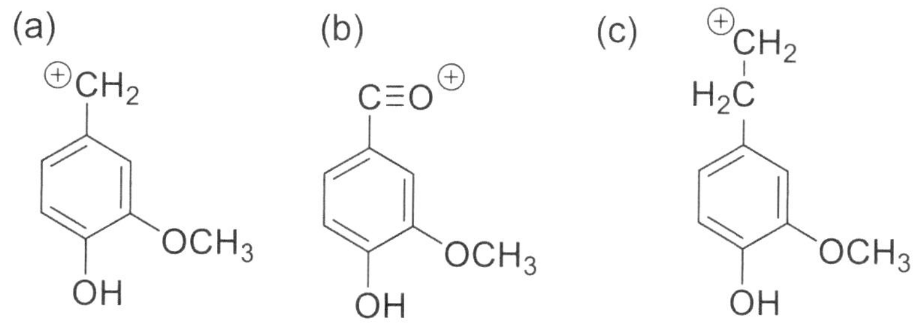


(a) $C_8H_9O_2^+$ (*m/z* 137.06) (b) $C_8H_7O_3^+$ (*m/z* 151.04) (c) $C_8H_{11}O_2^+$ (*m/z* 151.08)

**Fig. 2** Representative fragment ions derived from guaiacyl unit of lignin

secondary ions are detected [8]. It is possibly difficult to assign such smaller fragments to a specific polymer because they can be derived from other organic chemicals. The ratio of these smaller fragment ions can be affected by the original structures, and multivariate analyses are an upcoming approach for efficiently utilizing these smaller fragment ions.

Ions of $[C_5H_8O_4][H]^+$ (*m/z* 133) and $[C_5H_6O_3][H]^+$ (*m/z* 115) were dealt as representative secondary ions of hemicellulose [8, 9]; however, these ions were suggested to be attributed from other components, and they do not necessarily exhibit hemicellulose distribution [10]. Further study is needed for visualizing hemicellulose by TOF-SIMS.

Lignin has plural monomer units and interunit connections; hence, its mass spectrum is more complicated than that of cellulose. The representative fragment ions of lignin, attributed to the guaiacyl unit of lignin, are shown in Fig. 2. Secondary ions originated from guaiacyl, syringyl, and *p*-hydroxyphenyl units differ in *m/z* 30 for every methoxy group (H to $-O-CH_3$). These fragment ions can be generated by the cleavage of the weak linkage, such as ether bonds. It is difficult to cleave a strong linkage, such as 5–5′ bond between phenyl carbons. From these points, the exemplified secondary ions shown in Fig. 2 are not without generated from all types of lignin structures [11–14].

In the case of low-molecular weight chemicals, most of them form molecular ions, such as $[M]^+$, $[M+H]^+$, and $[M-H]^+$ (e.g., *2* and *7*). These characteristic ions can be easily interpreted in the mass spectrum using pure chemicals; however, there are various chemicals in biological samples, and the possibility of the overlapping of mass spectral peaks should be concerned.

### 3.3 Image Analysis

As shown in Fig. 3, the total ion image typically shows an image similar to a microscopic photograph, albeit each pixel provides different information. Making a selected ion image should be a simple analysis. After determining the characteristic secondary ion of the

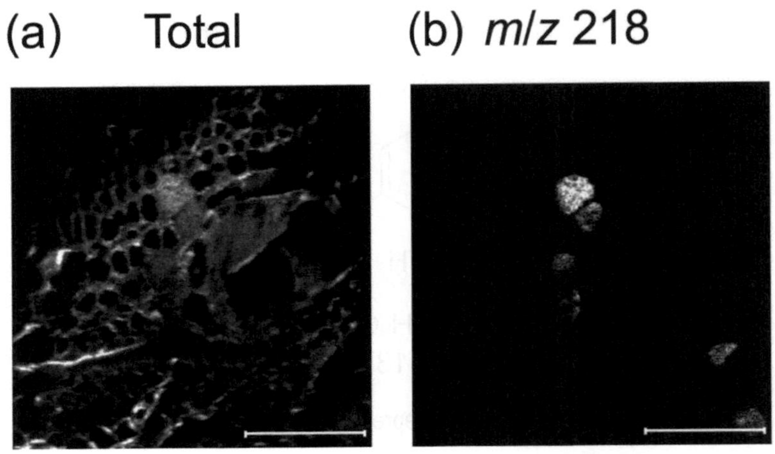


**Fig. 3** Example of TOF-SIMS (**a**) total ion image and (**b**) selected ion image for *m/z* 218. Scale bars are 100 μm

target chemical, its distribution in the measurement view field can be visualized (Fig. 3b). Ion intensities for each pixel are displayed using arbitrary color scales, such as binary, gray, thermal, and rainbow.

If there is a specific region or tissues in the view field, region of interest (ROI) analysis can provide a region-specific mass spectrum, which aids in the search of region-specific ions. The schematic illustration of ROI analysis is shown in Fig. 4. The measurement view field was divided into ROI1 (Fig. 4a) and ROI2 (Fig. 4c). Respective mass spectra for ROIs were extracted from the RAW data file. From the comparison of their mass spectra (Fig. 4b and d), it is suggested that ROI1 contains much more *m/z* 218 ion than ROI2.

### 3.4 Collaboration with Other Measurements

TOF-SIMS corresponds to microscopic mass spectrometry; however, its image resolution is less than that of electron microscopy. The actual amounts of the detected chemicals are not directly proportional to their secondary ion yields. Distributions of macromolecules are interpreted using their fragment ions. Furthermore, in several cases, it is difficult to identify a chemical in a biological system only by mass spectral information because of the possibility of the overlapping of mass spectral peaks. To overcome these limitations and exploit the potential of TOF-SIMS, combined analyses with microscopy, chromatography, spectroscopy, and other mass spectrometry are typically effective.

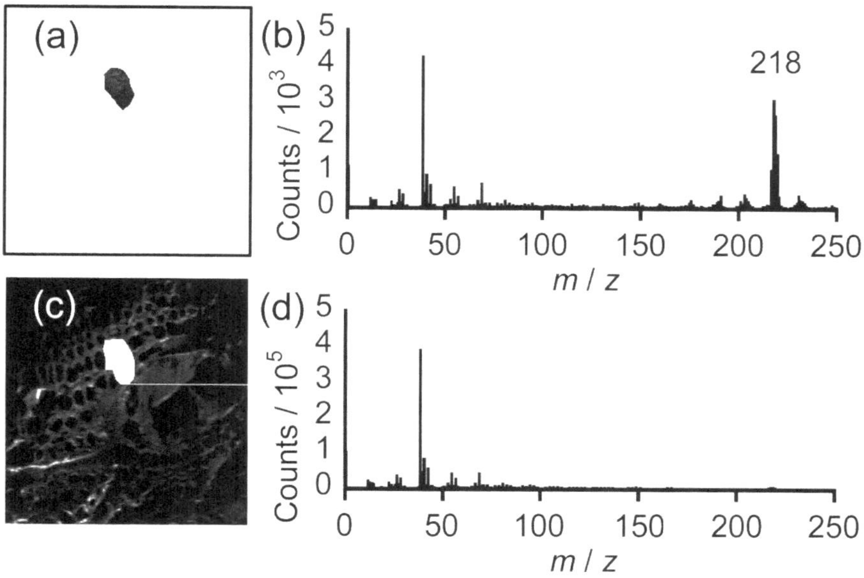


**Fig. 4** Example of ROI analysis. (**a**) Region and (**b**) mass spectrum of ROI1, and (**c**) region and (**d**) mass spectrum of ROI2

## 4 Notes

1. If the experimental target is composed of living and/or fragile tissues, such as the cambial zone and differentiating xylem, the procedures should be conducted with special precautions, such as those indicated in **item 3** of Subheading 2.1.
2. Wood material is an insulator, and TOF-SIMS measurement requires charge compensation as in the case of scanning electron microscopy (SEM) observations. Besides, TOF-SIMS measurement requires high-vacuum conditions. Oversized samples may require long decompression time, resulting in low measurement efficiency. The allowable thickness of the wood section varies according to the TOF-SIMS apparatus employed. The experiment performer should check with the TOF-SIMS operator. In most cases, a section of a few tens of micrometers is available.
3. For example, a carbon tape for SEM is available; however, the carbon tape can be a source of PDMS pollution; hence, it is better to fix it just before the measurement and use a small tape as far as possible. Gecko-inspired adhesive tape containing no gluing agents might be better if available.
4. As mentioned previously, the ionization behavior in TOF-SIMS measurements can differ according to the surrounding chemical and physical conditions. For example, pure cellulose powder and cellulose powder with inorganic salt result in different mass

spectra. Furthermore, the fragmentation behavior of polymer materials also changes according to conditions, such as crystallinity, polymer chain orientation, and measurement temperature. Typically, these effects are known as matrix effects. Measurements for standard chemicals, as well as the interpretation of results, should be conducted by taking into account their chemical and physical conditions.

## References


1. Vickerman JC, Briggs D (2001) ToF-SIMS: surface analysis by mass spectrometry. IM Pub. and SurfaceSpectra Ltd, Chichester
2. Kuroda K et al (2014) The accumulation pattern of ferruginol in the heartwood-forming *Cryptomeria japonica* xylem as determined by time-of-flight secondary ion mass spectrometry and quantity analysis. Ann Bot 113:1029–1036. doi:10.1093/aob/mcu028
3. Kuroda K et al (2013) The cryo-TOF-SIMS/SEM system for the analysis of the chemical distribution in freeze-fixed *Cryptomeria japonica* wood. Surf Interface Anal 45:215–219
4. Derue C et al (1999) A new cold stage for SIMS analysis and imaging of frozen-hydrated biological samples. J Trace Microprobe Tech 17:451–460
5. Dickinson M et al (2006) Dynamic SIMS analysis of cryo-prepared biological and geological specimens. Appl Surf Sci 252:6793–6796
6. Metzner R et al (2008) Imaging nutrient distributions in plant tissue using time-of-flight secondary ion mass spectrometry and scanning electron microscopy. Plant Physiol 147: 1774–1787
7. Aoki D et al (2016) Distribution of cell wall components by TOF-SIMS. In: Kim YS, Funada R, Singh AP (eds) Secondary xylem biology. Academic, London, pp 363–379
8. Fardim P, Durán N (2003) Modification of fibre surfaces during pulping and refining as analysed by SEM, XPS and ToF-SIMS. Colloids Surf A Physicochem Eng Aspects 223: 263–276
9. Tokareva EN et al (2011) Characteristic fragment ions from lignin and polysaccharides in ToF-SIMS. Wood Sci Technol 45: 767–785
10. Goacher RE et al (2011) Expanding the library of secondary ions that distinguish lignin and polysaccharides in time-of-flight secondary ion mass spectrometry analysis of wood. Anal Chem 83:804–812
11. Saito K et al (2005) Identifying the characteristic secondary ions of lignin polymer using ToF-SIMS. Biomacromolecules 6:678–683
12. Saito K et al (2005) A new analysis of the depolymerized fragments of lignin polymer using ToF-SIMS. Biomacromolecules 6: 2688–2696
13. Saito K et al (2006) A new analysis of the depolymerized fragments of lignin polymer in the plant cell walls using ToF-SIMS. Appl Surf Sci 252:6734–6737
14. Matsushita et al (2013) Fragmentation mechanism of the phenylcoumaran-type lignin model compound by ToF-SIMS. Holzforschung 67: 365–370

# Index



Miguel de Lucas and J. Peter Etchells (eds.), *Xylem: Methods and Protocols*, Methods in Molecular Biology, vol. 1544,
DOI 10.1007/978-1-4939-6722-3, © Springer Science+Business Media LLC 2017